DATA NETWORK HANDBOOK

An Interactive Guide to
Network Architecture and Operations

DATA NETWORK HANDBOOK

An Interactive Guide to

Network Architecture and Operations

by

Kenneth Reed

VAN NOSTRAND REINHOLD

I(T)P® A Division of International Thomson Publishing Inc.

New York • Albany • Bonn • Boston • Detroit • London • Madrid • Melbourne
Mexico City • Paris • San Francisco • Singapore • Tokyo • Toronto

Copyright © 1996 by Van Nostrand Reinhold

I(T)P ™ Van Nostrand Reinhold is a division of International Thomson Publishing, Inc.
The ITP logo is a trademark under license

Printed in the United States of America

For more information, contact:

Van Nostrand Reinhold
115 Fifth Avenue
New York, N.Y. 10003

International Thomson Publishing
GmbH
Königswinterer Strasse 418
353227 Bonn
Germany

International Thomson Publishing Europe
Berkshire House 168-173
High Holborn
London WCIV 7AA
England

International Thomson Publishing Asia
221 Henderson Road #05-10
Henderson Building
Singapore 0315

Thomas Nelson Australia
102 Dodds Street
South Melbourne, 3205
Victoria, Australia

International Thomson Publishing Japan
Hirakawacho Kyowa Building, 3F
2-2-1 Hirakawacho
Chiyoda-ku, 102 Tokyo
Japan

Nelson Canada
1120 Birchmount Road
Scarborough, Ontario
Canada M1K 5G4

International Thomson Editores
Seneca 53
Col. Polanco
11560 Mexico D.F. Mexico

1 2 3 4 5 6 7 8 9 10 QEBFF 02 01 00 99 98 97 96

Library of Congress Cataloging-in-Publication Data
Reed, Kenneth.
    Data network handbook: an interactive guide to network architecture/
    Includes index.
    ISBN 0-442-02299-9
    1. New business enterprises—Communication systems. 2. Internet advertising. 3. Interactive marketing. 4. Internet (Computer network).
I. Title
HD62.5.J36 1995
025.06'658—dc20
                                               95-45550
                                                  CIP

CONTENTS

PREFACE

The purpose of this book is to provide a conceptual view of computer networking. The primary audience for this book is the CIS major and the MIS professional. Having worked in the computer networking industry for over twelve years and having taught at two Universities in the Denver area for over three years, I have found that the approach presented in this book provides the basic understanding needed by both the student and the active networking professional.

Every detail of networking does not need to be understood to grasp how networks work. There are however some critical concepts, that once understood, provide a foundation that allow the student and professional to quickly grasp and apply new concepts and technologies in a practical way. Also provided with this book is a CD-ROM which gives additional detail on many of the subjects covered in this book. For instance, when studying MAC layer protocols such as Ethernet, the book covers how Ethernet works and common ways to implement Ethernet. All these subjects can also be found on the CD, along with additional subjects such as Ethernet traces, showing how Ethernet traffic looks on a real LAN. Also on the CD are tests and exercises that will further develop the concepts found in the book.

The book is arranged in four primary sections. The first section covers enough basics to understand general networking concepts and the remaining subjects found in the book. Additional concepts can be found on the CD which provide information on fundamental concepts and application of these concepts. A thorough on-line glossary can also be used which gives additional detail on these topics.

The second section covers local network architectures, beginning with local area network architectures and ending with network operating system architectures. Both subjects are logically tied together to explain how LANs transport data between nodes and how NOSs are used for sharing resources across these LANs. The most popular LAN and NOS architectures are covered in this section, with others to be found on the CD.

The third section covers system architectures. IBM's SNA is covered as well as DECNet and TCP/IP. These architectures exist simultaneously in many of our corporate networks and operate together. The fourth and final section discusses how the architectures, both PC network architectures and system architectures, can be tied together.

The fourth section covers the integration of architectures. The components that are used to tie homogenous and heterogenous networks together are covered. These include repeaters, hubs, switches, bridges, routers, and gateways. This section also covers the telecommunications links used to connect these architectures together. Finally, this section and the book close with a look at a typical network, demonstrating many of the concepts found throughout book.

1

FUNDAMENTAL CONCEPTS

CHAPTER OBJECTIVES

This first chapter will introduce you to some of the fundamental concepts which underlie the entire book. Models will be presented which will help you to better understand and apply the concepts and information presented throughout the book. The objectives of this chapter are to:

- Identify some of the characteristics of connectionless and connection-oriented networks

- Differentiate between a process, a service, and a protocol

- Explain the differences between a LAN, a MAN and a WAN

- Examine the different ways nodes can be connected to make networks

- Explain why protocols are arranged in layers

- List the ways in which two processes can cooperate in a layered architecture

- Define the effects of changes to layers within a layered architecture

- Characterize the differences between data communicated by the lower layers and by the upper layers

- Explain what addresses are and how different layers use them

- Explain the flow of data and data encapsulation in a layered architecture

1

NETWORKING CONNECTIVITY

Networks can be classified according to how nodes (computers) are connected: point-to-point channels, where two computers share a single communication channel, or broadcast channels, where many nodes share the same channel.

Point-to-Point Channels

Point-to-point channels are not shared by more than two nodes at one time (See Figure 1-1). Many nodes might be connected to a single channel, but one of the nodes on the channel is the master. The master allows no more than two nodes to use the channel at one time.

Figure 1-1. Point-to-Point Network Example

Point-to-point networks can be further broadly classified according to how nodes which are not directly connected communicate with one another. Point-to-point networks are either circuit-switched or packet-switched.

Circuit-switched networks establish a physical connection between two nodes, and a packet is passed between nodes by "switching" it through intermediate points, either through other nodes or through a host computer (See Figure 1-2). A packet is a bundle of data that might contain all or part of information exchanged between applications. Circuit-switched networks are analogous to the voice telephone system, and such connections are often termed virtual circuits.

The virtual circuit establishes a single route for data between the nodes that does not vary for the life of the connection. Therefore they are connection-oriented. IBM's SNA is an example of a circuit-switched network.

Packet-switched networks go to the other extreme. They are termed connectionless because no connection is established. Instead, packets of information are passed from node to node, with a packet possibly traversing many nodes before it arrives at its destination. Many packets can be moving between the same nodes at the same time. This is like the telegraph network of old.

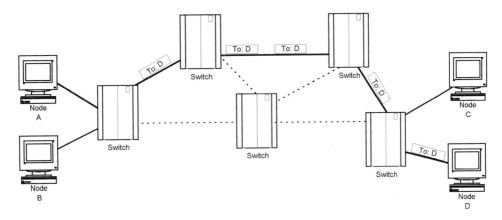

Figure 1-2. Circuit-Switched Network

Packets in a packet-switched network are analogously termed datagrams. A TCP/IP network is an example of a packet-switched network (See Figure 1-3).

   In the early days of data communications, all networks were circuit-switched, and many still are. More recently, for networks that cover wide areas, the emphasis has shifted to packet switching, simply because it permits the interconnection of far more nodes into a single network. With packet switching, fewer communication channels are required (because channels are shared by many users), and interconnection of networks is much easier to accomplish.

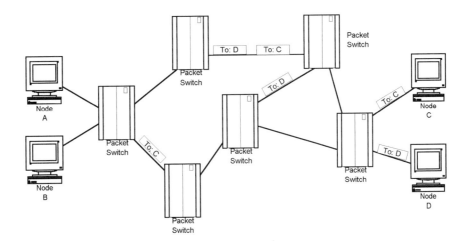

Figure 1-3. Packet-Switching Network

The layering of data communications protocols will be discussed later in this chapter. As you will see, layering gives rise to situations where the lower layers of a network are connectionless, but the higher layers establish a connection. The opposite can also occur, where the lower layers establish a connection and the upper layers do not.

Broadcast Channels

Broadcast channels are, as the name implies, similar to short-wave radio conversations. A node broadcasts on the common channel, and all other nodes connected to the channel listen to the message. The channel might consist of electromagnetic emissions (that is, radio), or a wire or cable to which all nodes are attached. As with users of CB radio channels, more than one node might begin to transmit at the same time. When that happens, each garbles the other's message. They all must stop and then start over again in such a way that subsequent collisions are not likely to happen. Alternatively, transmissions can be scheduled so that more than one transmission never takes place at the same time. You will learn in a later chapter about the protocols that have been developed to avoid or to handle these "collisions."

NETWORK EXTENSION

Networks can be classified according to the area over which they extend. A local area network (LAN) can consist of a few nodes or up to several hundred, but will typically be confined to one or a few buildings within a few thousand meters of one another. It can consist of subnetworks linked together in certain ways to form the larger, but still local, network. A subnetwork is a portion of a network in which all of the nodes are directly connected For example, all of the nodes may be connected by one piece of wire (See Figure 1-4).

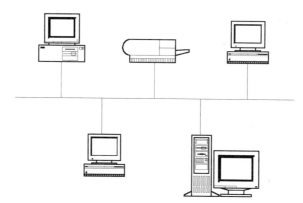

Figure 1-4. Subnetwork Example

Metropolitan area networks (MANs) are evolving and are being developed primarily by data carriers in response to the demand to interconnect LANs across a metropolitan area. For example, a university might interconnect its campuses. Wide area networks (WANs) are often interconnected LANs or MANs. They can be homogeneous, interconnecting like networks, but are often heterogeneous, that is, interconnecting LANs or MANs that have been built using different technologies. A WAN can span campuses, cities, states, or even continents. Typically, only one node on each LAN or MAN, called a router (or gateway), connects to the WAN. WANs are connected by a telecommunications link. Nodes on one LAN communicate with nodes on other LANs via the router or gateway (See Figure 1-5).

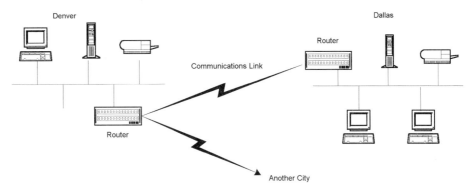

Figure 1-5. Wide Area Network Example

NETWORK TOPOLOGIES

A topology is a generalized geometric configuration of some class of objects that join together. With respect to networks, topologies describe different ways that nodes can be connected to make networks.

Point-to-point networks can have several different arrangements of the links. The five topologies that actually occur in real point-to-point networks are shown in Figure 1-6. The choice of topologies is often a matter of the technology being used for the network, or of geographical considerations. Each topology has advantages and disadvantages, which you will learn as you study the architectures that use them later in this book.

Most of the broadcast networks you will study in this book have a wire or cable as the communication channel, rather than using any form of radio communications. They can have two topologies: bus and ring. A network with a bus topology has a single wire or cable with multiple taps. The ends of the bus are unconnected. A network with a ring topology has a single wire or cable which connects to itself to form a closed loop (See Figure 1-7).

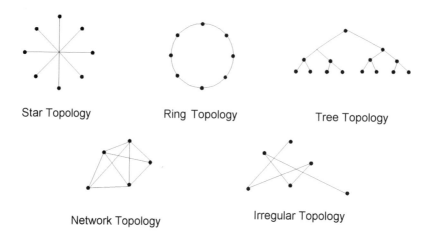

Star Topology   Ring Topology   Tree Topology

Network Topology   Irregular Topology

Figure 1-6. Point-to-Point Topologies

Note that point-to-point networks also can have ring topologies, but in that case, the nodes are connected by multiple point-to-point channels, rather than a single wire.

Although the ring topology for broadcast networks is discussed here for completeness, it will not be studied in this book because pure rings are uncommon in today's networks.

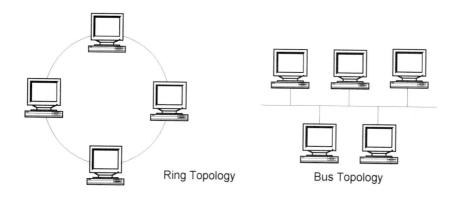

Ring Topology   Bus Topology

Figure 1-7. Broadcast Network Topologies

PROTOCOLS

Data communications involves the transfer of data between computer programs. Just as humans must share a common language in order to communicate, the programs must have a common protocol. The protocol simply defines the format and meaning of the data that the programs interchange.

An example of a very simple protocol might be the following (See Figure 1-8). Suppose two programs which are connected by a communications channel need to exchange messages which vary in length.

Figure 1-8. Simple Protocol Example

The protocol might specify that the first three characters of each message are numeric characters which give the length of the message itself (in decimal, not counting the first three characters). For the message "HELLO WORLD," which contains eleven bytes (including the space), the sending program would transmit the characters "O11HELLO WORLD." The receiving program would accept the first three characters, and then, knowing the length of the message, would expect eleven more characters.

If the receiving program got a message that had something other than numerics in the first four characters, it would consider that an error. If the sending program stopped sending before all of the characters were received, that would be considered an error, and so on.

PROGRAMS AND PROCESSES

The term "program" tends to mean a complete set of functions which provide a high-level function of some sort. Therefore, in the literature of data communications, and at some points in this book, the term "process" will be used instead of "program". Usually "Process" is used when referring to some subset of functions (still possibly quite complex) which fit into a larger program or is part of a large system, and especially when we are talking about a program when it is executing.

Types of Cooperative Processes

Processes that communicate with one another obviously must be cooperating in some sense to accomplish a useful function. But they can cooperate in different ways. Three important styles of cooperation are given names: peer-to-peer, client/server, and master/slave.

Peer-to-Peer Processes

Two programs or processes which use the same protocol to communicate and perform approximately the same function for their respective nodes are referred to as peer processes. With peer processes, in general, neither process controls the other, and the same protocol is used for data flowing in either direction. Communication between them is spoken of as "peer-to-peer." Later in this book, when you study layered communications protocols, you will see many examples of peer processes-communications using the same layer of the protocol stack (See Figure 1-9).

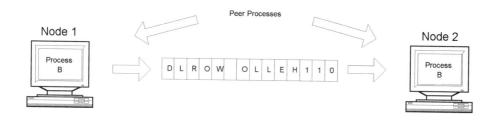

Figure 1-9. Peer-to-Peer Process

Client/Server Processes

Another way that processes can cooperate is for one process to take the role of client and the other that of server. The client process makes requests, via a shared protocol, for the server process to perform some task. This method of cooperation is typically used to allow the sharing of resources across a network (See Figure 1-10). The client process takes advantage of a resource that exists on the server node. Examples of resources that are commonly shared through client/server arrangements are computing cycles, graphics capabilities, file sharing, printing, and databases. Typically, the client process is found on a lower-capability, end user node, such as a workstation or PC, while the server process runs on nodes with larger capacity or greater power, such as a network file server. Examples of programs which employ client/server architectures are the Network File System (NFS) and relational database management products.

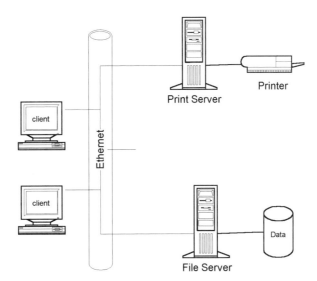

Figure 1-10. Client/Server Configuration

Both client and server processes are dedicated to their respective tasks and the roles do not ever reverse. The processes share a common protocol, but the protocol defines entirely different conventions for communications originating from the client than those originating from the server. This is differnet than peer-to-peer communications, where the protocol is more or less the same in both directions.

Master/Slave Processes

In a communications system that is hierarchical in arrangement, processes are often spoken of as being either masters or slaves. The issue is one of control rather than service. Master/slave relations often occur in cases where one node has greater "intelligence" than the other; that is, greater computing capacity. Master/slave relations occur in the IBM SNA environment because of its hierarchical nature. The protocol used in master/slave relations is generally the same for communication in either direction, but the master node is in complete control. An example is the relationship between a mainframe and a 3270 cluster controller (See Figure 1-11).

Protocol Converters

There is no such thing as a standard data format for data communications. Two programs that communicate with one another must always share the same protocol, unless another program or

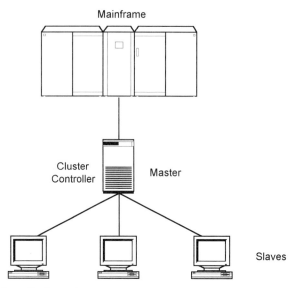

Figure 1-11. Master/Slave Configuration

a piece of hardware is interposed between them to translate from one protocol to another. Such a device or program, if transparently dedicated to this task only, is called a "protocol converter."

Services

If the program at the end of the communication link is an interactive program, then it is providing some service to the user; for example, accepting and displaying messages that are transmitted by the other program. The user may have no knowledge of the protocol that the programs are using to exchange data, so services are quite distinct from protocols. Many different protocols could be used to provide a given service, and many different services could be provided by programs which use a single protocol to communicate.

A communications program can provide its services to another program instead of to a user. In fact, when we talk about services in this book, they usually will be services provided by one program to another.

PROTOCOL LAYERING CONCEPTS

Early data communications programs were monolithic—that is, a single large program provided many services and communicated with other programs with a single "low-level" protocol (See

Figure 1-12). This did not work well. The programs were hard to change, and the protocols would not work well as new technology came along. This led to the development of protocol layering.

Figure 1-12. Monolithic Communications Programs

A layer is simply a program or set of programs which provides services to the next higher layer and uses services of the next lower layer. A program which resides at the highest layer will typically provide many sophisticated services to the user, but most of these services are actually implemented, directly and indirectly, by the lower layers.

Because a program provides services only to the layer above it and uses services only of the layer below it, a change to any given layer will affect only the layer above it. Layering breaks a single monolithic program into parts which are isolated from one another, making the program easier to write and to change. Layering does, however, extract a performance penalty. There is some overhead associated with moving data through multiple layers.

Protocol Layers

Layering applies to protocols as well as to services. In a system which has a layered architecture, a process communicates only with its peer processes. Otherwise, as with services, a change to one layer would affect many other layers. Peers communicate with a common protocol, appropriate to the services they provide. Therefore, each level requires a protocol that is different, so layers of processes have corresponding layers of protocols.

This concept of layering gives rise, in its simplest form, to two layers of services and two layers of protocols, as shown in Figure 1-13.

Layered Communications Systems

The simplest service/protocol model has two layers, but the ideas it illustrates can be extended to an arbitrary number of layers. In fact, real data communications systems typically involve from three to seven layers of services and protocols.

Figure 1-13. Layers and Services

Levels of Abstraction

The layers of data communications services can be viewed in a general way as shown in Figure 1-14. Programs at the lowest layer provide services related to the simplest, most "concrete" form of data: streams of bits. Programs at the highest layer provide services related to the most complex or "abstract" forms of data: data that are ready to be displayed to humans, program data structures, or, in the case of programs using more advanced object-oriented techniques, objects.

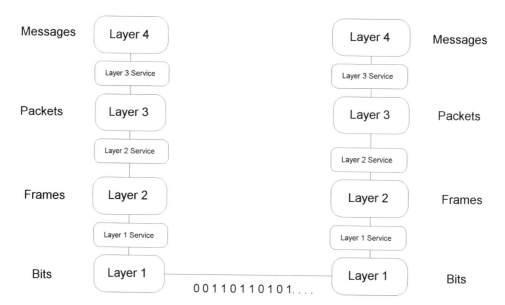

Figure 1-14. Levels of Abstraction

Programs at the intermediate layers transform the data from the simple to the complex forms, and vice versa. This transformation is sometimes referred to as "abstraction," and you might hear the phrase "levels of abstraction," referring to movement from the concrete to the abstract, that is, from the lowest level to the highest.

Layering and Routing

One reason that protocols vary from layer to layer is because the services offered by each layer to successive layers progress from concrete, bit-oriented services to more abstract, higher level services for data objects. But protocols differ for another important reason: the lower layers communicate across a single link, while the upper layers must communicate with peers to which the link is indirect, that is, through other nodes.

The bit stream oriented protocol at the lowest level supports simple streams of bits flowing between two points. Since the only language understood at this level is "0" and "1," the protocol recognizes only peers to which it is directly connected by the physical link across which the bits flow. It simply has no way of addressing anyone else, and that is not its job anyway.

If communications between peers at all levels were bound by this restriction, then communication possibilities would be extremely limited. But a program at a higher level can invoke services from more than one process on the lower level, and thus from more than one physical link. Note that the two lowest level protocols do not need to be identical protocols; in fact, often they will not be.

Since higher levels can pass more complex messages, they can include routing information, making it possible for data to flow across more than one communication link, and making large networks possible.

So an important service provided by lower level protocols in any of the networking systems we will study (which are all layered) is the ability to handle the routing of messages between computers which are not connected by a physical link.

In a system with a layered architecture, programs at a high level deal with more complicated data but not with implementation details and routine tasks. For example, in a network; the routing of messages is done by lower levels, so upper level programs need not be aware of it. The data flows only through the highest level that is concerned with handling them. So when a message flows through an intermediate node, the data will flow only through the lower level processes involved with their transmission.

Protocol Stacks

Layers of protocols are often called "protocol stacks" or "protocol suites." As with services, the lowest level protocol deals with bits. The highest level protocol deals with complete data structures or objects that are ready to be used by an application program or displayed to a user.

Protocols and Addresses

An address in data communications terminology is a number used to identify a destination in a network. Addresses used at different layers of a protocol stack identify different points in a network. As you will see later in this book, addresses are used to identify networks, physical nodes attached to a network, as well as processes within a node on a network.

ENCAPSULATION AND DECAPSULATION

Data communications programs accept data from the layer above them and pass data, after possibly having transformed them in some way, to the layer below. They must usually communicate with their peer program in some way, using the protocol established for their level, in order to perform their functions. For example, suppose a program divides a large file into shorter segments. It must somehow tell the receiving program how many segments it is sending and in what order they should be reassembled. One possible alternative is to modify the data, but that would require that the sender provide some portion of the data for the use of lower levels. That would make even the highest levels sensitive to the operation of the lower levels — not a good solution.

Virtually all modern communications systems solve the problem this way: the intermediate levels simply add prefixes, called headers, and sometimes postfixes, called trailers, to the data. This is called encapsulation (See Figure 1-15).

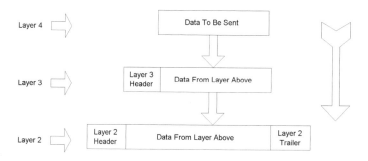

Figure 1-15. Data Encapsulation

The header normally contains a field specifying the length of the encapsulated data together with at least one field providing information about the data. If, for example, the data is a segment of a large file, the header might specify the relative position of the segment in the complete file and probably the total number of segments that make up the file.

The receiving peer program removes the added information (decapsulation) after processing the data, but before passing the data on to the higher layers. The original file is not disturbed, and the higher layers never see the encapsulating information. See Figure 1-16.

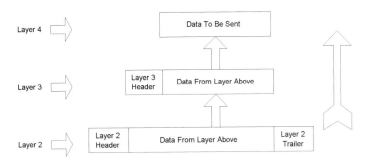

Figure 1-16. Decapsulation

SUMMARY

A network connects nodes, some of which are hosts to which terminal nodes attach, in two different ways: point-to-point and broadcast. Point-to-point networks fall into two classes: circuit-switched networks, in which a connection is formed between the nodes, as in a telephone network; and packet-switched or connectionless networks, in which packets of data, or datagrams, are passed from node to node until they reach their destination, like telegrams.

All of the nodes in a broadcast network share the same channel, and the network protocol controls access to the channel and avoids or recovers from the collisions that occur when more than one node tries to use the channel at the same time.

LANs, local area networks, extend across a single site and consist of one or more subnets, which are usually, but not necessarily, homogeneous. MANs, the metropolitan area networks are being developed by data carriers to connect LANs in the same city. WANs are wide area networks, often heterogeneous, which cover many sites, spanning large corporations and sometimes continents.

Five common topologies exist for point-to-point networks: star, ring, net, tree, and irregular. For broadcast networks which do not use a radio band, two topologies have been developed: ring, where the ends of the channel are connected, and bus, where they are not.

Anything or anybody that communicates must share a common protocol with that with which it communicates. Programs that communicate with one another using the same protocol are called peers. Data communications programs are layered. Peer programs at each layer provide service to the next higher layer, and use the services of the next lower layer to perform their assigned function.

Programs in the lower layers provide services which are less abstract and which relate to moving data from one place to another. Programs in the higher layers are concerned with more abstract functions, such as presenting data in various forms. The layering of programs gives rise to protocol stacks, which define a data communications standard.

Protocols use addresses to identify specific locations in networks and processes within nodes on individual networks. These addresses are located in specific fields of protocol headers.

Understanding data communications will be easier if you understand the differences between services and protocols. With that foundation, you can approach the study of data communications programs in terms of the services they provide and the protocols they use to provide those services.

ADDITIONAL INFORMATION ON THE CD-ROM

The CD-ROM contains additional information on subjects found in this chapter. Use the "Search" function to view the following subjects:

- Networking basics section
- Networking basics tests and exercises
- Additional information on terms and concepts in this chapter

2

THE OSI MODEL LAYERS 1-4

CHAPTER OBJECTIVES

This second chapter will introduce you to the Open Systems Interconnection (OSI) Model. The objectives of this chapter are:

- Name the lower layers of the OSI model and give their relative position in the stack

- For the physical, data link, network, and transport layers, you should be able to:

    - describe the service that the layer provides to the layer above it;

    - name some of the common protocols used by each layer; and

    - identify the function and benefit of the multiplexing and parallelization provided by the transport layer.

OVERVIEW OF THE OSI MODEL

Programs must use a common protocol to communicate. If many different protocols for data communication exist, then the potential for linking computers into common networks is obviously limited. Networking protocols were first developed by computer manufacturers, with each manufacturer developing its own protocols. Even individual manufacturers had more than one protocol, because protocols were developed independently for different computer platforms. IBM, for example, had more than a dozen protocols back in the 1960s.

This chaotic situation has led to the creation of standards—both official and de facto. The de facto standards—TCP/IP for UNIX users, SNA for IBM users, and DECnet for Digital users—will be covered in Chapters 6-8.

Building on the concept of layered protocols, this chapter will introduce the very important Open Systems Interconnection (OSI) model. OSI began as a reference model; that is, an abstract model for data communications, but now the model has been implemented and is in use for data communications. The OSI model is used to provide a logical structure to the course and will serve as a reference throughout the remainder of the chapters.

The OSI model, consisting of seven layers, falls logically into two parts. Layers one through four, the "lower" layers, are concerned with the communication of raw data. Layers five through seven, the "higher" layers, are concerned with the networking of applications. This chapter discusses the four lower layers, and the following chapter covers the three higher layers. It is important that you study OSI for two reasons:

1. OSI provides a general model of a layered internetworking architecture which is more detailed and specific than the service/protocol model we discussed in the first chapter; and

2. The OSI model is widely used in current literature. Recent texts and periodicals on data communication make use of the OSI model as a means to structure the presentation of their material and explain functions of products.

OSI shows the direction in which networking is moving, setting the stage for the de facto standards you will study.

The OSI model was created by the ISO, the International Organization for Standardization. It is composed of members from the national standards organizations of eighty-nine countries, including ANSI, the principal non government U.S. standards organization.

Note: ISO is not an acronym. The organization's name was borrowed from the Greek word "isos," meaning equal. Isos is also the root of the prefix "iso" (as in isometric) meaning equal dimensions. Since the organization has three official languages — English, Russian, and French, creating an acronym would be difficult.

The OSI standard was patterned after and is similar to, but simpler than, the IBM layered networking scheme, Systems Network Architecture (SNA). SNA was introduced by IBM in the early 1970s to provide a consistent standard for all of its product platforms.

In Figure 2-1 you see the OSI reference model. It consists of seven layers of services and protocols. The term "open" is used in the name because the intent of the standard is to allow the interconnection of networks without regard to the underlying hardware, as long as the communications software they use adheres to the standard. However, vendors of data communications products have not been "open" to this idea.

The layers of OSI provide the levels of abstraction mentioned earlier; from the concrete Physical Layer, which deals with bits, to the abstract Application Layer, where user applications, such as an electronic mail interface, would reside.

We will now look briefly at each of the layers, from the lowest to the highest, to learn about the services and protocols at each level.

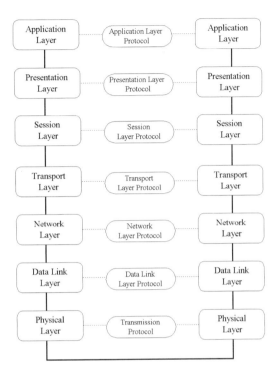

Figure 2-1. The OSI Model

THE PHYSICAL LAYER

The Physical Layer is the lowest layer in the OSI reference model. The Physical Layer is responsible for the transmission of bits across a communications channel, which can range from a coaxial cable, fiber optic cable, or satellite link to an ordinary telephone wire (See Figure 2-2).

The Physical Layer processes provide the service of transferring bits across the physical link. This is done without knowledge of the meaning or structure of the bits. They do not know, for example, whether they are transferring 8-bit bytes or 7-bit octal characters. Some faults, such as an open connection, can be detected and the error indication passed on to the higher layers, but most error detection and all error correction are the responsibility of higher layers.

The transmission protocol that the processes of the Physical Layer use vary according to the nature of the link. The protocol is concerned with such things as:

- how bits are represented;
- how to tell when transmission starts and ends; and
- whether bits can flow in one direction only or in both directions at the same time.

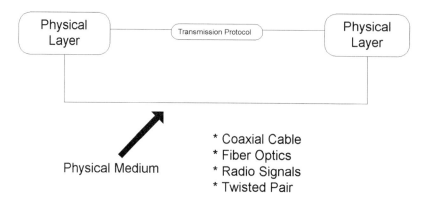

Figure 2-2. The Physical Layer

These all depend on the type of communications channel being used, with the characteristics of twisted pair phone lines being quite different from those of, say, a coaxial cable. An example of a physical layer encoding scheme would be Manchester encoding. Examples of transmission protocols are Ethernet, Token Ring, and RS-232. Figure 2-3 shows several examples of physical layer devices. You will learn more about the physical layer details when you study local area networks in Chapter 4.

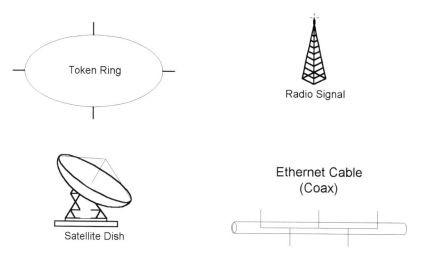

Figure 2-3. Physical Layer Transmission Mechanisms

THE DATA LINK LAYER

The Data Link Layer is the second layer in the OSI reference model. The Data Link Layer is concerned with the transmission of frames, rather than just bits, across the physical link. A frame is a series of bits which form a single unit of information understood by communicating Data Link processes. (See Figure 2-4).

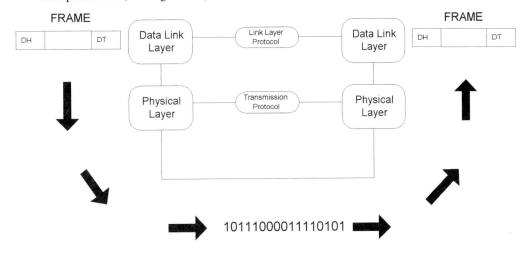

Figure 2-4. Data Link Layer

The Data Link Layer provides two classes of service:

- connectionless, unacknowledged service, or

- connection-oriented, acknowledged service.

Connectionless, unacknowledged service can be provided when the data link is very reliable and higher layers are handling error correction. For connectionless service, this layer has little to do besides break the stream of data from the upper layers into frames, as explained below. Providing connection-oriented, acknowledged service involves a great deal more, so, for the remainder of this section, connection-oriented service will be discussed.

At the transmitting end, the Data Link Layer provides the following services to the layer above—the Network Layer:

- It accepts the address of an adjacent node to which it is to transmit the data. An adjacent node is one to which a physical layer link exists.

- It accepts data packets of arbitrary length from the network layer.

- It handshakes with its peer to ensure that the complete frame is received correctly.

At the receiving end, the peer process participates in the handshaking and passes the received packets up to the Network Layer.

To provide these services, the transmitting Data Link Layer must:

- Divide the packets into frames.

- Put sequence information into the frames. If they get out of sequence during error recovery, they can be put back into the right sequence by the receiving peer process.

- Add error detection and correction codes to the data frames, so the receiving peer process can tell when an error has occurred.

- Add handshaking (control) information to the data frames, so it can cooperate with the peer process to correct problems, such as a frame that is completely lost.

- Ensure that it does not send frames at a faster rate than the receiving process can handle them.

- Use the services of the Physical Layer to transmit the frames of the packet. The peer process on the receiving end must do the reverse.

- Check the frames for errors, taking appropriate corrective measures when errors are found (such as requesting a retransmission from the transmitting peer process).

- Put the frames back into the correct sequence to reconstruct the packet.

- Handle problems, such as missing frames, through handshaking with its peer process.

The Data Link Layer does not have complete responsibility for error detection; varying degrees of error correction can occur in higher layers.

Several bit-oriented protocols will be studied later in this book. Common bit-oriented data link protocols used by this layer are:

- High-Level Data Link Control (HDLC), an ISO standard, and subsets such as:

    1. Synchronous Data Link Control (SDLC), an IBM protocol

    2. Link Access Protocol B (LAPB),used for X.25 networks

3. Link Access Protocol D (LAPD), used in ISDN networks

4. IEEE 802.2 Logical Link Control (LLC)

Data Link Layer Summary

The Data Link Layer is the lowest layer which can establish a connection and ensure the error-free transmission of data. It establishes a connection (if necessary), accepts packets of data from the Network Layer, divides them into frames, and uses the Physical Layer to transmit them. At the other end, it reconstructs the packets and performs necessary error recovery, including requesting the retransmission of missing frames and throwing away duplicate frames.

THE NETWORK LAYER

The Network Layer is the third layer in the OSI reference model. The Network Layer handles all of the problems associated with getting a packet of information from one node to another in a network when the message must pass through a third node because the source and destination nodes are not directly connected. That third node is called an intermediate node or a hop.

A process in the Data Link Layer communicates with only one other peer process: the one at the other end of its communications link. But a process in the Network Layer communicates with peer processes at the other ends of all communications links to which the node attaches. The job of the Network Layer is to send and receive packets (See Figure 2-5). If the node is an intermediate node, the Network Layer in that node is responsible for forwarding packets onward to their destination (See Figure 2-6). It must handle packets to and from different node types, which might use different communications protocols and different addressing schemes.

The Network Layer lies at the border between subnets and full networks. This layer provides the following services to the Transport Layer:

- A unified addressing scheme, so that each node has a unique address and the addressing scheme for the whole network is consistent. This may mean resolving disparate addressing conventions and duplicated node addresses between differing types and versions of subnets

- Establishment and maintenance of the virtual circuit for circuit-switched networks

- For packet switched networks independent routing of each packet through each intermediate node

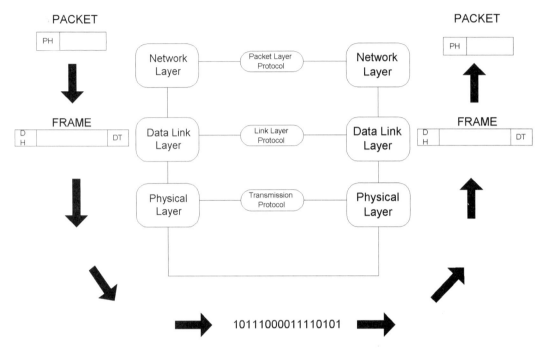

Figure 2-5. The Network Layer

Circuit switched networks establish virtual circuits between pairs of nodes, across which all data transfer takes place in both directions. You will recall that circuit-switching is analogous to voice service for today's phone system. Routing is established when the connection is made (you dial the number once at the start of the phone call). Individual packets need not contain addresses, because routing has been established and intermediate nodes know in advance where to send each message (after you dial the number and are connected, you don't have to dial again). When an intermediate node fails, the circuit is broken and must be reestablished through some other route before data transfer can resume (when a phone line goes down, it must be fixed, unless the phone company has another line across which it can route the call).

Since the virtual circuit is dedicated to the connection, service is predictable and will not vary much due to other traffic in the network (when you are talking on the phone, it seems like you have a dedicated wire connecting you to the other person), but when the traffic is light on the virtual circuit, circuit capacity is potentially wasted (toll charges add up even if you are not talking).

Packet-switched networks on the other hand, are analogous to the telegraph service. Each packet must contain the full network address of the destination node (you have to put the person's name and address into the telegram).

Each intermediate node must decide where to transfer the packet next (the telegraph operator at each station must look at the message and decide where to send it next).

When a node goes down, the network can immediately reroute messages along an alternate

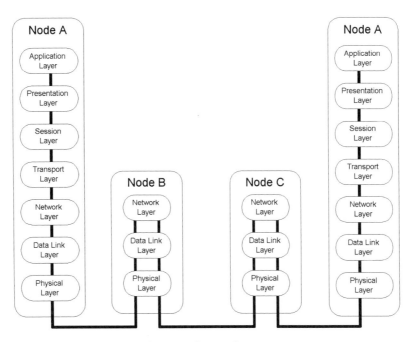

Figure 2-6. Communication through intermediate nodes

route (when the line goes down between Topeka and Dallas, messages to Topeka can be routed through Wichita instead).

Since circuits are shared, traffic can impact service (the telegraph operator can get too busy to send your message), but data links can be more fully utilized (telegraph operators can send other messages while you write your reply).

Common protocols used by the Network Layer include:

- X.25, a connection-oriented, packet-switching protocol defined by the ITU-T. It is widely used for public data network, especially in Europe

- Internet Protocol (IP), one of the networking protocols developed for the DARPA Internet project

Network Layer Summary

Processes in the Data Link Layer deal only with a single adjacent node. Processes in the Network Layer deal with two or more peer processes, taking responsibility in connectionless networks for routing a packet received from one neighbor onward to the next, until the packet reaches its final destination. In connection-oriented networks, processes take responsibility for routing a packet according to the established connection. The Network Layer programs must be able to communicate with heterogeneous peers and must provide consistent addressing of nodes across the network.

THE TRANSPORT LAYER

The Transport Layer is the fourth layer in the OSI model. As you learned in the previous section, in the Network Layer, peer processes are always on adjacent nodes; their view of the world is limited. The Transport Layer sees, in a manner of speaking, the whole network, using the lower layers to provide "end-to-end" communication for the higher levels.

The job of the Transport Layer is to "transport" messages from one end of the network to the other (See Figure 2-7). It is the lowest end-to-end layer, that is, the lowest layer in which the peer processes at either end of the connection carry on a conversation using their common protocol. Processes in the Transport Layer act as if the nodes were adjacent, relying on the lower

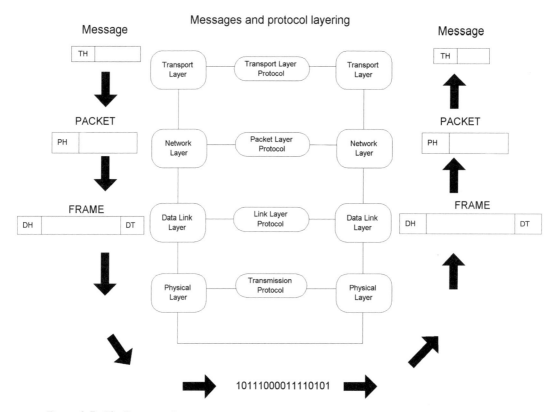

Figure 2-7. The Transport Layer

levels to take care of passing data through intermediate nodes to get them across the network.

The Transport Layer has been termed a "keystone" for the layer hierarchy, because it frequently resides at the border between the delivery mechanism: for example, a public X.25 network, and the enterprise network.

Basic services provided by the Transport Layer include addressing, connection management, flow control, buffering, and resource allocation.

Addressing

The Transport Layer takes responsibility for making connections to specific processes within a node. All of the lower layers need only concern themselves with network addresses—one address per node. But there might be many processes on a given node which are communicating at the same time. For example, two users might be transferring files across the same link at the same time (multiplexed, of course). The Transport Layer handles addressing of processes on the node.

Connection Management

The Transport Layer is responsible for establishing and releasing connections, which is more complicated than one might think, because of the chance of lost and duplicate packets.

Flow Control and Buffering

Each node on a network is capable of receiving messages at a certain rate. That rate is determined by the computing capacity of its computer and other factors. Each node also has a certain amount of processor memory available for buffering data. The Transport Layer is responsible for ensuring that sufficient buffers are available in the receiving node and that data are not transmitted at a rate that exceeds the rate at which the receiving node can accept them.

Resource Utilization

Another job of the Transport Layer is to match the resources available through the lower levels to the requirements of the upper levels. It does this through multiplexing and parallelization. When the upper layers require slower service than the available channel can provide, the Transport Layer will multiplex conversations across a single channel to make more effective use of the channel's bandwidth. It does this by interleaving the packets of different messages. The Transport Layer process at the receiving end then sorts the packets, recreating the original messages based on the user ids stored in the Transport Layer headers (See Figure 2-8).

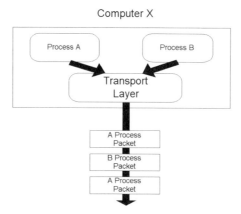

Figure 2-8. Transport Layer Multiplexing

When the upper layers require faster service than a single channel can provide, the Transport Layer will combine multiple channels, paralleling the flow of data across them to multiply the effective throughput rate for the higher layers (See Figure 2-9). This assumes that the Transport Layerprocess is running on a machine which is capable of multiprogramming, so that the lower

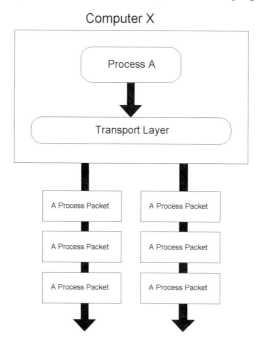

Figure 2-9. Transport Layer Parallelization

layer processes which support the actual data transfer across the channels can operate in parallel. Multiprogramming capability is a function of the operating system used on the nodes. Operating systems such as UNIX, and mainframe operating systems such as MVS, provide multiprogramming, while some PC operating systems do not.

Service Quality

The Transport Layer also has responsibility for the quality of the communication service provided to the session layer. For some networks, any or all of several errors might occur in the Network Layer. Messages might be:

- Corrupted

- Lost

- Delayed for an inordinate amount of time

- Delivered out of sequence. The corollary to this is that duplicate messages might be delivered out of sequence

- Delayed

The OSI Reference Model lets the user of the Transport Layer specify, quantitatively, the quality of service required. This is done by providing Quality of Service (QOS) parameters along with the request to establish a communication channel or to deliver a packet. Parameters are defined for both circuit-switched (connection-oriented) and packet-switched (connectionless) service.

The QOS parameters support different levels of service based on the requirements of an application. For example, an interactive application used for inquiries to a text database might specify very high QOS parameter values for connection establishment delay, throughput, transit delay, and connection priority, to get good response time, while specifying a lower QOS value for residual error rate/probability, because the impact of errors on the user is relatively low (the user just requests a retransmission). A file transfer application used to move updated data around a network at night, on the other hand, can accept slower response, but must have reliable, error free transfer of data, so correspondingly different values might be specified.

The following protocols are commonly used in the Transport Layer:

- ISO 8073. This ISO standard was developed for use in networks implementing the OSI model. It defines a protocol for use by programs which implement the services described in this chapter.

- Transport Control Protocol (TCP). It was developed, with the Internet Protocol (IP), mentioned earlier, for the Internet.

Transport Layer Summary

The Transport Layer is at the heart of the OSI protocol stack, falling at the border between public and private networks, providing end-to-end communication. It must provide reliable, cost-effective data transmission, in spite of a possibly unreliable network underneath it, and it insulates the higher levels from all concerns about the transportation of data.

The Transport Layer must manage connections, provide flow control and buffering, handle the multiplexing and paralleling of channels, and manage the network to meet user-specified service quality parameters.

SUMMARY

The job of these four layers is to move data reliably across the network. The Physical Layer is concerned with moving bits across a link and maintaining a physical connection.

The Data Link Layer is responsible for dividing packets into frames and moving the frames across a given physical link, keeping them in sequence, and recovering from the kinds of errors it is able to detect, such as transmission errors. But it knows only of its peer at the other end of the link. The Data Link layer address is concerned with getting a frame of data from one point in a network to the *next point* in the network.

The Network Layer knows of several peer processes—one for each data link—and takes responsibility for moving packets between links to sending them on their way to their final destination. The Network Layer address is concerned with getting a single datagram or packet from one end of a network to the *final destination*.

The Transport Layer is aware of the entire network and is the first end-to-end layer. Its peer is located at the final destination of the packet and it relies on the Network Layer to move packets across intermediate nodes. A transport layer address is concerned with getting a complete message from the process on one end of a network to the *peer process* at the final destination.

Figure 2-10 demonstrates how these four layers work together to communicate information across a network. Process A running in Node 1 on Network X wishes to transmit data to a peer process located on another network (Y). Addresses used at each layer are used to get the information from process to process. The Transport Layer address refers to the destination of the receiving peer process (A). The Network Layer address refers to the destination of the receiving network (Y). The Data Link Layer address refers to the hardware address of the receiving node (2). All addresses work together to communicate data from process to process. These concepts will be expanded upon in later chapters.

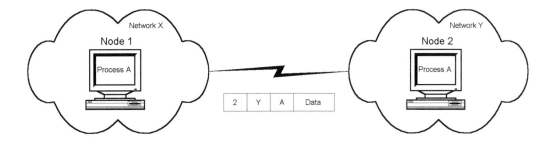

Figure 2-10. OSI Layers 1-4 Functionality

ADDITIONAL INFORMATION ON THE CD-ROM

- OSI Model test questions
- Additional information on terms and concepts covered in this chapter
- Examples of physical layer standards such as Manchester Encoding
- Categories of physical cables and cable specifications
- Traces of data link, network, and transport layer protocols
- Examples of packets and frames
- Examples of service qualities found in various protocols

3

THE OSI MODEL LAYERS 5-7

CHAPTER OBJECTIVES

This chapter will introduce the final three layers of the OSI model. Important differences exist between the four lower layers you have studied so far, as a group, and the remaining three layers.

The job of the lower layers is to provide reliable communication of data across a network. The job of the upper layers, taken collectively, is to provide user-oriented services through a set of widely available standard applications and through applications written for the users by programmers.

The job of the upper layers taken individually is to provide services that ease the task that application programmers face in writing the applications that reside in the Application Layer. Therefore, as you will see, these layers are substantively different from the lower layers.

To conclude these first three introductory chapters, you will be introduced to three widely used network protocols: the standard for most UNIX-based network platforms, TCP/IP, IBM's SNA, and Digital Equipment Corporation's DNA (DECnet). We will describe their relationship to the conceptual models we have talked about. These three network architectures are studied in more detail in later chapters. At the completion of this chapter, you should be able to:

- Explain how the lower four layers, taken collectively, differ from the upper three layers of the OSI model

- Name common protocols used by the Session, Presentation, and Application Layers

- Describe the services that each layer provides to the layer above them

- Name the important network applications and briefly describe the services provided by each

- State important characteristics that TCP/IP, SNA, and DNA (DECnet) have in common or that set them apart from one another

THE SESSION LAYER

The Session Layer is so named because it deals with the notion of a session taking place between two entities; for example, an interactive user session would begin with the user logging on to the computer and end with the user logging off. A session is different from a connection established by the Transport Layer because of its longevity, as explained below, and because the Session Layer provides services that go beyond simply establishing a connection.

A session is analogous to a conversation between humans. Certain conventions in conversation allow for the orderly and complete transfer of information between the parties:

- They first agree to talk to one another

- They don't talk at the same time

- They divide the conversation into parts ("Let me describe it to you, and you can tell me what you think.")

- They end the conversation in an orderly fashion ("I'll talk to you later." "Okay. Bye.")

Similarly, the Session Layer provides the higher layers with services that can be invoked to conduct sessions, including:

- Establishing a session (separately from a connection)

- Conducting dialogs (prevent both parties from transmitting data at the same time)

- Managing activities (divide the session into parts)

- Ending the session gracefully (both ends agree to stop)

You might wonder about the application's access to the Session Layer, given that another layer, the Presentation Layer, lies between the Session Layer and the Application Layer. Can the application go around the Presentation Layer? No. The Presentation Layer simply passes the services of the Session Layer through to the Application Layer by including them in the suite of services it provides. All the Presentation Layer has to do to satisfy a request for a Session Layer service is pass it through. Obviously, some overhead results from this passing through of requests, but the process maintains the separation between the layers of the architecture. A change at the Session Layer would be isolated from the applications.

Establishing, Conducting, and Ending a Session

A session can be independent from a specific Transport Layer connection (See Figure 3-1).

There are two possibilities:

(A) Several sessions can take place during a single connection, or
(B) A session can require several Transport Layer connections to complete.

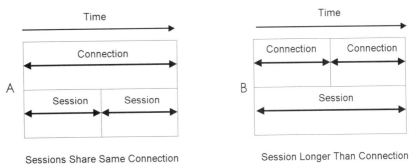

Figure 3-1. Sessions and Connections

In case A, several sessions can take place without the overhead of establishing the connection again and again. In case B, a connection can be interrupted and reestablished without disturbing the session. Note, however, that the Session Layer will not multiplex several sessions onto a single Transport Layer connection. That is the job of the Transport Layer.

In comparison to a human conversation, Case B is equivalent to calling the other party back if the call is interrupted before you have finished talking. Case A amounts to passing the phone around in a family call.

An important part of the Session Layer services is the "orderly release" of the connection. The lower layers support only an abrupt termination of the connection. In a conversation, it's polite to make sure that the other party is finished talking before you hang up the phone. The Session Layer takes care of this for dialogs between nodes.

Dialogs

The Transport Layer allows simultaneous communication in both directions across a channel (full duplex), but conversations require a dialog, where only one party speaks at a time.

The Session Layer provides a service which allows the application to conduct a dialog; when one node is "talking," the other is "listening." This is managed with "token passing," the mechanics for which are part of the service provided by the layer (See Figure 3-2). Only the node which has the token can talk. The nodes pass the token back and forth so the dialog can take place.

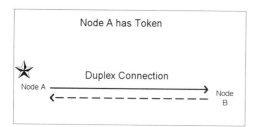
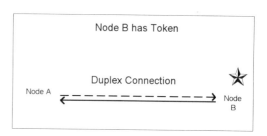

Figure 3-2. Session Layer Dialogs

Activity Management

With the preceding services, the application can start and stop conversations and alternate speakers. It is also necessary to identify where data starts and ends as it flows between the nodes participating in a conversation. For example, if several files are being transferred within a file transfer application, the receiving node must be told where one file ends and the next starts. Delimiters could be inserted into the data stream by the sender, of course, but then the recipient has to scan the stream for them, which requires computing cycles. Data that look like delimiters can also cause problems.

Activity management solves these problems. It divides the data stream into "activities." For the file transfer example, each file becomes an activity. The Session Layer inserts control information to mark activities in the header and signals the start and end of activities to the application.

Within activities, the Session Layer supports "sync points." Sync points amount roughly to saying "did you get that?" in a human conversation where you have been providing information that the other party is writing down. You make sure they have finished writing before you continue talking.

Sync points do a similar thing for programs. A sync point means "get everything I've sent you processed before you go on." Taking the file transfer application as an example again, a sync point might mean, "make sure that all the data I've sent you are written to disk before you take any more." Then, if something happens to disturb the flow of data after that point, both ends of the connection are confident that they can start over from that point.

Common protocols used by the Session Layer include ISO 8327, defined for use by programs being written to conform to the OSI model, the Advanced Program to Program Communication (APPC) facility of IBM's SNA, and the session control protocol of Digital's Digital Network Architecture (DNA).

Session Layer Summary

The Session Layer is the lowest layer in OSI that is not concerned solely with the transmission of data. It can map sessions onto consecutive Transport Layer connections or conduct consecutive sessions on a single Transport Layer connection. It also provides other services, including facilities for coordination of dialogs and management and synchronization of activities.

THE PRESENTATION LAYER

The Presentation Layer handles the representation of data as they flow between nodes. The lower layers provide the service of transferring data between nodes in an orderly fashion and ensuring that what is received is what has been sent. The Presentation Layer provides services that relate to the way data are represented.

Data Representation

The Presentation Layer resolves differences in the way data are represented in different computers attached to the network. For example, it handles communication between an IBM mainframe which uses the EBCDIC (Extended Binary Coded Decimal Interchange Code) character coding and an IBM or compatible PC that uses the ASCII (American Standard Code for Information Interchange) character code.

Data Security

The Presentation Layer encrypts and decrypts data so that anyone who covertly accesses the communication channel cannot obtain confidential information, alter information as it is being transferred, or insert false messages into the stream. It authenticates the source of information, that is, confirms that a party to a communications session is indeed the party represented.

Data Compression

The Presentation Layer also represents data during transmission in a compact fashion in order to make optimal use of the channel. It does that by compressing data passed to it by the Application Layer and decompressing the data before passing them back at the receiving end.

Problems with Data Representation

Data are represented in different ways on different computers: a computer cannot process data in the form used by another computer when the forms are different. The primary ways in which data vary are:

Character Coding

EBCDIC and ASCII are by far the most common codes, but others exist as well.

Byte Ordering Within Integers

The first (left-most) byte of an integer can be either most significant or least significant. For IBM mainframes, the Apple Macintosh and the Sun Microsystems SPARC models, bytes within integers are ordered in an intuitive manner, with the most significant digits to the left, so that the binary number 1 would appear as "00000001" in processor memory. In machines such as the IBM PC and others based on the Intel microprocessors, and in Digital Equipment Corporation products, the least significant bit is first, so 1 would be "l0000000."

Format of Floating Point Numbers

A floating point number consists of a mantissa field, an exponent field, and a sign. Each is represented with some number of bits in a certain order within a word (single precision) or a double-length word (double precision). The number of bits for each varies from computer to computer.

Furthermore, the way in which data are represented in certain compilers varies, even for compilers which compile the same language. For example, a Boolean value (a variable which takes two values, "true" and "false") might be stored as a byte on one computer but occupy a word (4 bytes) on another.

OSI Data Representation

Differences in data representation could, of course, be handled by the application program, and in many networks they are. But the Presentation Layer of the OSI model provides a generalized way of dealing with them called Abstract Syntax Notation 1, or ASN.1 With ASN.1, the burden of data translation is removed from the application programmer's shoulders.

ASN.1 provides a standardized format for data transfer between nodes (See Figure 3-3). Each node is only concerned with translating to and from ASN.1. There is no need to know anything about the format in which data are stored elsewhere on the network.

ASN.1 data transmission proceeds as follows:

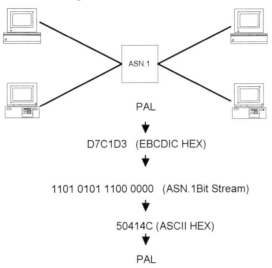

PAL

⬇

D7C1D3 (EBCDIC HEX)

⬇

1101 0101 1100 0000   (ASN.1 Bit Stream)

⬇

50414C (ASCII HEX)

⬇

PAL

Figure 3-3. Abstract Syntax Notation

1. The application programmer must write, in a syntax similar to PASCAL programming syntax, a definition of the program data structures.
2. The application program provides the ASN.1 definition along with the actual data structure to the Presentation Layer process.
3. The Presentation Layer process makes the necessary conversion of the data from native format (for example, EBCDIC) to ASN.1 format and returns the structure into a self-defining bit stream; that is, a bit stream wherein each data element in the stream is preceded with a code that defines its type and length. The bit stream format is called transfer syntax.
4. The data are transmitted via the lower layers.
5. The Presentation Layer process at the receiving end uses the corresponding ASN.1 definition (its name is included in the bit stream) to convert the bit stream back into an application data structure in the proper format for the application at that end.

Data Security

The security of data within the network falls within the province of the Presentation Layer. Three threats must be guarded against:

- Unauthorized use of the network, including false identification. False identification means pretending to be someone else, either by logging onto the network as that person or sending messages that are falsely attributed to another user of the network.

- Stealing data from the network (wiretapping, etc.).

- Inserting false messages into the stream, or removing messages from the stream.

The first threat is related to user authentication, which often is the responsibility of the logon process of the operating system. Operating Systems (or extensions of operating systems, like IBM's TSO) have a variety of means of detecting attempts to bridge their user authentication. For example, most systems will not allow more than a few attempts at entering a password. This makes the breaking of passwords by repetitive experimentation more difficult.

When user authentication is the responsibility of the application, the application must provide user authentication safeguards similar to those that the operating systems provide. These safeguards lie within the province of the application rather than that of the network.

Theft of data and insertion of false messages are network issues. The threat of both can be minimized by encryption of messages at their source and decryption at their destination. The application header data that authenticate and sequence the messages must be encrypted, as well as the data themselves, to prevent false messages or deletion of messages.

A wide variety of methods for data encryption exist. In general, they involve using a "key" to either reorder the bits, bytes, or words of a message (a transposition cipher), or substituting one or more encryption bytes or words for the "plaintext" bytes or words of the message (a substitution cipher).

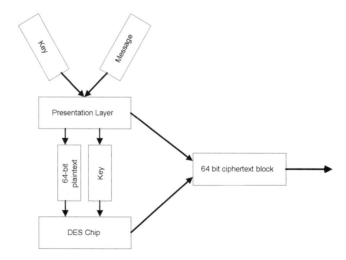

Figure 3-4. Data Encryption

The National Institute of Standards Technology (NIST, formerly the National Bureau of Standards) has adopted a standard for data encryption originally developed by IBM. Called the Data Encryption Standard (DES), it defines a method of encrypting data in 64-bit blocks using a 56-bit key. A complex algorithm with nineteen distinct stages, DES encryption uses the same key for encryption and decryption, essentially running the algorithm backwards for decryption (See Figure 3-4).

DES key and data blocks are long enough that DES encoded data are impossible to decode without having the key. The only way to decode the message without the key is by trial and error. Given the number of trials that would be required to discover the key and the amount of computation required for each trial, it would take the fastest computer many years to discover a single key.

Because DES is a standard, and because it is quite compute-intensive to encode and decode data using the DES algorithm, it has been embedded in silicon (a specific chip). As shown, the DES chip accepts a 64-bit block of plaintext and the key; it then outputs the ciphertext. The same chip can be used to reverse the process by inputting the ciphertext and the key to produce the plaintext.

Data Compression

Data transmitted between nodes of a network can often be quite repetitive. Financial data, for example, often contains long sequences of zeros. Many messages contain far more printable characters than unprintable characters, and more blanks, vowels, and numeric characters than consonants and signs. Since the amount of data actually transmitted between nodes in a network ultimately governs the cost of operating the network and the network's capacity to do useful work, it may be desirable to "compress" data before transmitting them. Data compression reduces the number of bytes that must be transmitted by translating the data in some way to a form that is more efficient and requires less storage.

If data are compressed by the sender, they must be expanded by the recipient, so what is needed is, of course, a protocol. A variety of data compression protocols have been defined. For example, a technique called "run length encoding" codes a sequence of many bits or bytes as a "count" field giving the number of bits or bytes in the stream followed by the bit or byte that is being encoded, as illustrated in the diagram. In this simple example, the "+" character is used to mark the start of a length-encoded field; naturally, that character cannot appear elsewhere in the input message, or the recipient will mistake it for the start of a compressed field. This problem is avoided by coding each "+" in the source message as "++" as shown in Figure 3-5.

Run-length encoding takes advantage of data that have repetitive adjacent elements. A number of other techniques take advantage of other characteristics of data. The decision as to which technique to use will depend on the specific application and the characteristics of the data.

Figure 3-5. Data Compression

Summary of the Presentation Layer

The Presentation Layer would be better named the "representation" layer. It is concerned with all of the problems of the representation of data that are transmitted between nodes of a network.

The OSI ASN.1 standard, by defining a neutral data format, takes care of the translation of one computer's data into the form of any other computer that supports ASN.1, avoiding the necessity of a translation function for every possible combination of source and destination.

Network security hinges on two factors: the security of the system from use by unauthorized persons, and the ability to encrypt data for transmission on the network. User authentication is the responsibility of the computer operating system or the application itself. The DES provides a standard method of encrypting and decrypting data, and special purpose integrated circuits have been developed for DES to off-load the host computer.

Data compression potentially reduces the traffic in a network through the use of protocols which take advantage of the characteristics of various types of data to compress them. This requires fewer bits to be transmitted, therefore increasing the capacity of the network.

THE APPLICATION LAYER

The Application Layer, the highest layer in the protocol stack, contains the applications that invoke the services of the network to get useful work done. These applications can be specified and written by programmers for purposes unique to the individual network, or they can be based on a more general tool, such as a database system that has been adapted by the user for specific needs. Network users have certain application needs that are quite common; for example, most users want to use electronic mail for day-to-day communication with colleagues.

Many frequently used applications are acquired from others, from the manufacturers of equipment used in the network, or from companies that sell such software.

Types of Applications

User-written application programs must have their own protocol in order to communicate. The OSI Application Layer process simply passes data from the Application to the Presentation Layer without modifying or encapsulating them.

If the Application Layer contained only custom user programs with their own custom protocols, we wouldn't need to say much about this layer. In fact, in the original OSI Model, the Application Layer was not "feature-rich." Over the years, requirements for services that needed to be provided at this layer (that is, services that didn't fit logically into the lower layers) were identified, and a number of Application Layer facilities were developed for OSI.

The Application Layer programs fall into two classes:

1. User applications which have not only a standard protocol of their own (at the Application Layer level), but also standard services that they provide directly to the user—in other words, OSI application programs. Chief among these is electronic mail (e-mail).

2. Application services, that is, programs which don't provide services directly to the user, but rather to the user-written programs. The purpose of these facilities is to keep application programmers from having to reinvent the wheel when they write network applications. Chief among these services is virtual filestores.

User Applications

Electronic Mail

E-mail is perhaps the most visible application on many networks. Its primary benefit is quick delivery of messages (almost instantaneously, in the case of messages to addressees on the same local area network).

The OSI standard for electronic mail is Message-Oriented Text Interchange Systems (MOTIS), which is derived from, and similar to, the X.400 standard developed by ITU-T, the United Nations authority concerned with telephone and data communications systems. MOTIS is defined in two parts: the User Agent, which is a definition of services provided, and the Message Transfer Agent, which is concerned with the protocol used to transmit the messages (See Figure 3-6).

The user agent might be called by a mail reader program, which adds a user-friendly front-end to the system. The user agent takes care of telling the user what's in the mailbox, accepting new messages, etc.

The message transfer agent implements the protocol for getting the message from the sender to the recipient. MOTIS, like virtually all e-mail systems, is a store-and-forward system, so it does not need to establish a connection to the recipient. The mail message simply becomes a datagram.

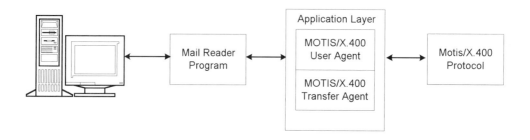

Figure 3-6. Message-Oriented Text Interchange

The ARPANET TCP/IP E-mail protocol, which is very widely used, is called Simple Mail Transfer Protocol (SMTP).

News (Network Bulletin Board)

News is closely related to e-mail; in fact, USENET, the most widely used network bulletin board, can be considered to be another layer on top of e-mail, because it uses e-mail to disseminate its postings. However, the SMTP protocol was designed to accommodate both news and e-mail, so news doesn't have its own protocol (See Figure 3-7). No news facility has been defined for OSI as yet.

A news group is a file of e-mail messages relating to a certain topic. For example, one USENET news group, called comp.databases, focuses on databases. You subscribe to the groups that interest you, and there are hundreds of them from which to choose. A group can be local (your site), regional, or international. Programmers or administrators of databases at a site might choose to establish their own local database group to address their own special interests.

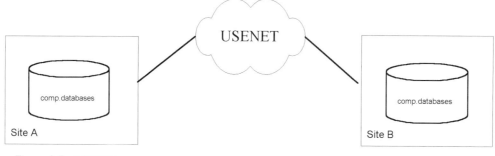

Figure 3-7. USENET

You access news through a special news reading program. The program initially notifies you of the groups that are available and lets you choose the groups to which you wish to "subscribe". The news reader then notifies you of when news arrives in your subscription groups and keeps track of which messages in those groups you have read, so you don't see them again (unless you ask to).

Each site maintains a copy of each of the news groups to which there are subscribers at that site. New messages for each group are promulgated from site to site by simply sending them as e-mail messages. If someone posts a message to a worldwide group, for example, the message is first put into the copy of the group at that site; then it is propagated from site to site until, in a short while, it has spread around the world.

File Transfer & Access

Two data manipulation capabilities are universally required by network users: the ability to copy a file from one node to another, called file transfer, and the ability to share a file that resides on another node, called file sharing or remote file access.

File transfer is straightforward: the user issues commands specifying the source and destination, and the file transfer facility transfers the file, leaving the original intact and creating a copy on the target node. Of course, the data might need to be transformed if the source and destination nodes are different. In the context of the OSI model, this is taken care of by lower levels (as you saw in the preceding discussion of the Presentation Layer). In the case of other environments, such as TCP/IP, the application itself must take care of it.

File Transfer Protocol (FTP) is used to transfer a file between two computers. UNIX systems also provide a widely used program called UNIX to UNIX Copy program, or uucp. Uucp uses its own special purpose protocol to copy files from one UNIX system to another and implement the USENET network.

File sharing is more complex than file transfer, because it is necessary to maintain only one copy of the file. When the remote node accesses a file in order to change it, the file, or at least the relevant portion of the file, must be "locked." Locking prevents another program from changing the same data at the same time, because those changes will be lost when the data are restored from the remote node. It is difficult to provide file sharing while preserving the integrity of the files and taking care of glitches in the network.

File sharing protocols allow the user of a node, as well as application programs which run on that node, to view the shared file as an extension of the file system on that node. The shared file looks and behaves as if it were local, except for differences in the speed of access to the data in the file. Transfer of data across a network is always much slower than transfer of data from a disk attached to the node.

The OSI File Transfer, Access, and Management (FTAM) model provide a "virtual filestore" to allow file sharing across a network. A virtual filestore is a file system that implements file sharing. To both the user and to application programs, the remote file appears to be resident on the node.

The Network File System (NFS) is another virtual filestore which provides the same sort of sharing as FTAM. It is a very commonly used system in TCP/IP environments, and you will see it demonstrated in the next chapter.

Virtual Terminals

No standard exists for terminal equipment. Terminals made by different companies require different control codes in the input streams that are sent to them in order to produce a desired result (even different models from the same vendor can require different codes). If the application program generates the control codes for a specific terminal, then it must be modified in order to use the application with any other terminal type—obviously not a very desirable situation.

As networks developed, many different terminals were in use, so to avoid the problem of incompatibilities, the concept of the virtual terminal was developed. Virtual terminal protocols provide an abstract definition of a terminal that is specific in terms of the control codes required to produce a specific result that is also general in nature (See Figure 3-8). The virtual terminal definition combines all of the common features of most real terminals but leaves out the "bells and whistles" that make terminals unique to their vendor.

Application programs are written to generate the control codes for the virtual terminals. A mapping function "driver" is then written for each type of actual terminal to translate the standard virtual terminal codes into the codes required by that hardware and to otherwise modify the stream of data to the device to conform to its standards. If the application needs to take advantage of any unique hardware on a given terminal, it must "escape" the virtual terminal protocol to issue the special commands—but then it may become incompatible with other terminal types.

The most commonly used virtual terminal is the TCP/IP TELNET definition. It can be argued that the IBM 3270 protocol has become a de facto virtual terminal protocol.

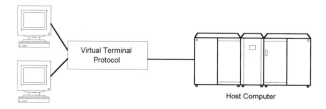

Figure 3-8. Virtual Terminal Protocol

Virtual Terminals Versus Terminal Emulators

We must be careful to distinguish terminal emulator programs from virtual terminal protocols. A terminal emulator program is a program which allows a computer, typically a PC, to emulate a terminal by mimicking the protocol of the terminal.

An example of a terminal emulator is a program called tn3270. It is available for the IBM PC and the Apple Macintosh. It makes the PC appear to the network to be a specific model of an IBM 3270 display device. Some terminal emulators emulate terminals which have the ability to communicate over phone lines, so the emulator also provides the ability to connect a remote PC to a node. For example, the tn3270 program emulates a model of the 3270 which is connected to the CPU through a network controller, rather than directly to the CPU's local input/output channel.

Confusion between terminal emulators and virtual terminal protocols arises because terminal emulators can emulate virtual terminals. For example, programs are available for the PC and the Macintosh which make them appear to be TELNET terminals. As a result, the term TELNET is used loosely by many to speak of the emulator program rather than the protocol.

Job Transfer and Management (JTM)

One of the earliest uses of data communications, in the days when most computing was initiated through "batch jobs" rather than interactively, was "remote job entry," or RJE as it is called by IBM. In those days, batch jobs were initiated with decks of punched cards read into the computer through a card reader. The results were printed on a line printer or punched into cards. Both card reader/punch and line printer were attached to the channels of the mainframe and could be no more than 400 feet from the mainframe. An ME device was a special purpose terminal: it combined a card reader/punch and line printer, their associated controllers, and a small, special purpose computer which controlled the devices and handled communications. It could communicate with the mainframe via phone lines, providing users with a remote facility for reading and punching cards and for printing. Cards could be read into the remote mainframe to initiate batch jobs, and the resulting output of the jobs, in the form of punched card images or print lines, would be punched or printed on the RJE device.

Job Transfer and Management (JTM) is the modern-day remote job entry facility. It is defined by the OSI standard as part of the OSI Application Layer. It provides the services necessary to execute a program on a specified remote node, supply the program with necessary input files (possibly obtained from other nodes), and dispose of the results as directed (possibly on some other node).

Application Layer Summary

Although many Application Layer programs are unique, user-written programs, a number of standard protocols have been defined and corresponding applications have been written. Application programs include user applications, such as e-mail, and application services, such as virtual filestores.

E-mail is one of the most widely used applications. The OSI MOTIS protocol is essentially identical to the X.400 specification. Network bulletin boards, such as the widely used USENET News, use SMTP to propagate news items as e-mail messages.

Data files can be both transferred (copied) between nodes on a network and shared (one copy of the data used by multiple nodes). The File Transfer Protocol (FTP) of TCP/IP and the uucp protocol used for USENET are examples of transfer protocols. Files are shared across a network by making it appear that a file resides on a local disk, although in fact it continues to reside on the disk of the node that owns it. Such a facility is called a virtual filestore. Sun Microsystem's Network File System (NFS) and OSI's File Transfer, Access, and Management (FTAM) provide this capability.

Virtual terminal protocols allow a wide variety of terminals to be used with one version of a program, because the program is written to support a virtual terminal standard protocol (such as the TCP/IP TELNET protocol) rather than being written to support a specific hardware device. The IBM 3270 protocol has become a de facto virtual terminal protocol because it is so widespread. Virtual terminal protocols are sometimes confused with terminal emulators, which mimic either real or virtual terminals.

SUMMARY

This discussion of the three highest layers of the OSI model, the Session, Presentation, and Application Layers, completes the presentation of the OSI Model. The job of the uppermost three layers is to provide services to the applications which reside in the highest layer. By providing a suite of protocols and a set of services to accomplish tasks that would have to be recoded over and over again, these upper layers provide standardization so that applications can more easily share data and communicate with one another. Taken together, layers five through seven provide capabilities for:

• Handling differences in data representation between computers

• Handling differences in physical characteristics of terminals on the network

• Ensuring that data are not stolen or compromised during network transfer and that the network is not used to gain unauthorized access to data

- Making the most efficient use of network resources

- Managing dialogs and activities by the application and synchronization of application activities

- Sharing data between nodes on a network

Additionally, a number of common network applications are widely available such as electronic mail, network news, file sharing, file transfer, virtual terminal definitions, job transfer and management.

SUMMARY OF THE OSI MODEL LAYERS 1-7

Figure 3-9 summarizes several important concepts studied in the first three chapters. As a frame is being built, each of the layers adds its own header to the data from the layer above (encapsulation). Note that the Data Link layer is the only layer that truly encapsulates the data by adding both a header and a trailer when constructing a frame. At the receiving end of the communications link, each layer removes its header (decapsulation) and passes the data portion up to the next layer. Peer layer communication occurs since corresponding layers process only the data that is contained in the specific layer header.

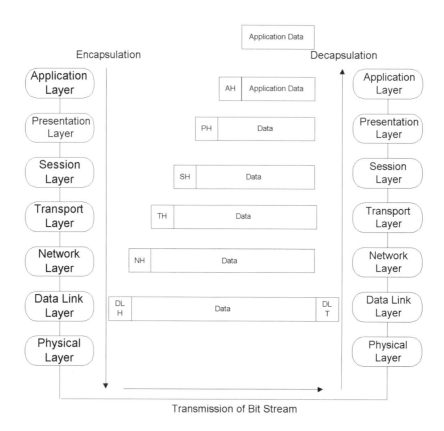

Figure 3-9. OSI Model Summary

ADDITIONAL INFORMATION ON THE CD-ROM

- OSI model test questions
- Additional information on terms and concepts covered in this chapter
- Traces of session, presentation, and application layer protocols
- ASCII and EBCDIC code tables
- Binary, hexidecimal, and octal code tables

4

LAN ARCHITECTURES

CHAPTER OBJECTIVES

The previous chapters were meant to provide the reader with an overview of networking and give a theoretical perspective of networking. By focusing on the OSI Model and the ideas and theories that underlie it, the chapters have provided a great deal of fundamental knowledge of networking technology. In this and the following chapters, the focus shifts to the many specific technologies and protocols that you will encounter using, analyzing, and implementing networks.

As with the early chapters, we will start with the lowest level in the protocol stack. It makes great sense to study Local Area Networks (LANs) at this point, because the architectures we will study in the later chapters all support the LAN technology we will learn about in this chapter. In fact, the majority of network users today are connected to their network via a LAN.

This chapter will teach you about the LANs that you are most likely to encounter in your businesses: Ethernet and Token Ring. We will also study FDDI, which is being used increasingly for backbone networks.

At the completion of this chapter, you should be able to:

- Identify important characteristics of baseband and broadband transmission and give the important characteristics of the baseband standards

- Contrast the characteristics of bus, ring, and star ring topologies

- Identify the components of the IEEE 802 protocol suite, showing their relationship to one another

- Explain why collisions occur in a CSMA/CD network

- Explain CSMA/CD message transmission logic

- Give the types and number of tokens that circulate on a Token Ring

- Explain how tokens are inserted onto and removed from a Token Ring and Token Bus

- Explain why Token Ring and Token Bus exhibit better performance characteristics than CSMA/CD under conditions of heavy load

- Give the rationale for FDDI's dual rings

- State the maximum throughput of the Ethernet, Token Ring, Token Bus, and FDDI LAN Standards

- Given a user's requirements for a LAN, identify the LAN architecture that addresses those needs

- Understand the frame formats used for the LAN topics discussed

HISTORY OF LANS

Data Communications began in the 1960s with an emphasis on communication over long distances—between different cities and often over much greater distances, but at relatively slow data transfer rates. At that time, that was what was needed. Most computing was done in batch mode, and companies needed an efficient way to transmit data processing files between computer sites. In the 1970s, interactive terminals began to proliferate, but most data communications took place over long distances—for example, processing transactions between a computer center in the suburbs and interactive terminals downtown.

Ethernet Development

In 1972, at Xerox's Palo Alto Research Center (PARC) in California, small computers began to be configured as "personal" computers, and their users began sharing resources, such as printers, across communication links which used the RS-232 protocol within a local area.

Although RS-232 can be used to connect two computers directly, it was developed for what are now called Wide Area Networks (WANs). It was intended to connect computers through modems and phone lines. RS-232 requires a thick cable for each computer, can handle data transfer rates of only a few thousand bytes per second, and can span distances of only a few hundred feet. It was not at all suitable for local area communications because it was too short, too slow, required too much wiring, was too expensive, and too difficult to administer. So researchers at Parc invented what became Ethernet in 1980 (See Figure 4-1). Ethernet ultimately became the IEEE 802.3 CSMA/CD standard in 1985.

Ethernet was designed to link computers by connecting them to a common cable (called a "bus"), in order to minimize the amount of wiring required, and to operate at a much higher speed than RS-232, with a burst rate of 10 megabits per second.

Token Ring Developed By IBM

IBM was also connecting small computers in a local area during the 1970s, but the computers were of a different sort—they were intelligent controllers for devices such as bank teller machines, where each machine had its own controller.

IBM developed a technique for connecting such machines which, like Ethernet, minimized the wiring required. But, rather than a bus, IBM connected the machines in a "ring," with point-to-point links between nodes on the ring, so that each node talked to only two others. The machines communicated by passing "tokens" around the ring, so the scheme was called a token ring.

As IBM PCs and compatibles proliferated in the early 1980s, Ethernet, by then a licensable technology and an emerging de facto standard, was used to interconnect them into LANs. IBM chose to adapt its token ring technology to compete with Ethernet as a means of creating LANs and in 1984 announced Token Ring for its PCs and workstations.

Token Ring initially operated at a burst rate of four megabits per second. In 1985k, IEEE defined the 802.5 Token Ring standard, essentially the IBM Token Ring. Today, 802.3 (CSMA/CD-Ethernet) and Token Ring compete as LAN standards. Each has its advantages and disadvantages. You will learn them later in this chapter.

IEEE Develops Token Bus for LANs in the Factory

As small computers grew in numbers in the office, they were also spreading to the factory floor, where they were being used for such tasks as controlling milling machines. Again, there was a need to interconnect these systems so that they could share

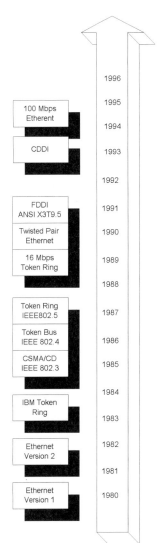

Figure 4-1. LAN Time line

common databases, such as numerical control programs for the milling machines.

Ethernet has many characteristics desirable for this environment. A failure of one Ethernet node on the network does not affect other nodes—one certainly would not want a whole factory to stop whenever a node fails. But Ethernet has one serious drawback. For reasons you will learn later in this chapter, the protocol cannot guarantee that a given node will be able to transmit data within a given time when the network is heavily loaded. Certain shop floor applications, such as guiding a robot which moves pieces from one manufacturing station to another, have "real time" requirements.

What was needed was a LAN technology that had a bus topology like Ethernet but that could guarantee that a node will always have access to the network, even when the LAN is heavily loaded. This is an advantage of token passing, where every node gets a chance to transmit periodically, no matter how busy the network is.

To satisfy the requirements of LANs on the factory floor, the IEEE combined the desirable elements of Ethernet and Token Ring, and the IEEE 802.4 Token Bus standard was born. Token Bus networks operate at 10 megabits per second.

Cheaper LANs, Faster LANs

Over time, the technology used to transmit and receive signals on the bus or ring improved significantly. The speed of Token Ring data transfer increased from its initial 4 megabits per second in 1985 to 16 megabits per second in 1989. The initial LANs used relatively expensive, shielded cables in order to achieve the high data rates required. By 1989, improved technology allowed "twisted pair" cables, that is, ordinary telephone wire, to be used instead. Twisted pair is much less expensive and easier to install than shielded cables. In some cases, it is even possible to use wire previously installed for telephones, avoiding the cost of installation altogether.

As LAN use increased, it quickly became desirable to interconnect LANs, to make each LAN a "subnet" of a larger network. The links between subnets require much higher data transfer rates than an ordinary LAN because of the large number of nodes ultimately connected to the network. This requirement led to the development of fiber optic technology for networking. Fiber optic signal transmission can provide much higher data transmission rates and was adapted to LAN use with the announcement of the Fiber Distributed Data Interface (FDDI) by the American National Standards Institute in 1990 (as ANSI X3T9.5). FDDI uses a ring topology and employs a token passing protocol, similar to 802.5 Token Ring, to provide a data rate of 100 megabits per second.

FDDI now appears in many networks as vendors bring more and more equipment based on the standard to market. But signaling technology for shielded copper wire has continued to improve to the point where FDDI data transmission rates have been achieved that are based on shielded copper wire. The X3T9.5 standard was extended to define a new standard with a copper-based medium called CDDI-Copper Distributed Data Interface. CDDI is a version of FDDI which runs on shielded and unshielded twisted pair cable.

Ethernet at 100Mbps is gaining some momentum as a replacement for 10Mbps Ethernet in situations where higher data transfer rates are needed. There are two new Ethernet standards which support 100Mbps transfer rates: IEEE Fast Ethernet and 100VG-AnyLAN.

TRANSMISSION METHODS AND MEDIA

The remainder of this chapter will focus on the technology used to implement LANs. LAN technology centers on three areas:

- The transmission method and medium used to transmit signals between nodes

- The topology of the network (bus, ring, star) and the tradeoffs between them

- The protocols used to transmit data between nodes on the LAN and for medium access control

For bus networks, medium access control means preventing or handling collisions; for ring networks, it refers to the method used for inserting messages on and removing them from the ring.

The two methods used to transmit signals between nodes are baseband and broadband (See Figure 4-2). In the context of LANs, broadband refers to analog transmission of digital signals, and baseband refers to digital transmission of digital signals.

Baseband is analogous to a telegraph. Zero voltage levels are modulated onto a constant carrier signal. Transitions from one level to the other indicate a "0" or a "1." Baseband is relatively simple and less costly than broadband, yet it is still very fast. It is far more widely used than broadband.

Broadband is analogous to TV cable transmission. Just as many different TV channels can broadcast different programs at the same time, a broadband link can support many independent communications channels. These channels can be used as independent links between nodes, or they can be used in parallel to increase the effective throughput of the link. Though potentially

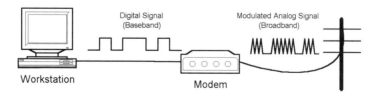

Figure 4-2. Baseband and Broadband Signals

much faster and able to span longer distances than baseband, broadband requires a modem at each end of a link, increasing the cost of every device attached to the LAN.

Fiber Optic Cable

Fiber optic cable consists of a light-transmitting core surrounded by a reflective cladding layer. Information is transmitted by focusing a light source into the core and then switching it on and off (See Figure 4-3). Light is reflected off of the cladding layer, so nearly all of the light radiated into one end of the cable reaches the other end. Fiber optic cable is not susceptible to loss of signal strength over distance, as is copper cable. And, since light is not appreciably affected by electromagnetic fields, a signal in a fiber optic cable is not affected by interference. But, by its very nature, a fiber optic cable must be cut in order to get the received light beam to a detector, as only point-to-point channels are possible; a bus cannot be constructed.

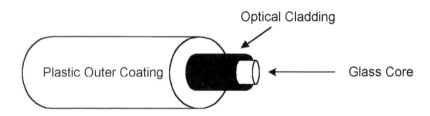

Figure 4-3. Fiber Cable

Twisted Pair

A twisted pair line, as the name implies, consists of a pair of wires twisted together (See Figure 4-4). Because the wires are twisted, electrical interference tends to affect both wires equally, so it does not affect the difference in potential between the two wires. This makes twisted pair cable less susceptible to signal loss than if it weren't twisted. Twisted pair lines are suitable for bit rates up to 950Mbps over short distances (less than 100m) and lower bit rates over longer distances. More sophisticated driver and receiver circuits enable similar or even higher rates to be achieved over much longer distances. These are referred to as UTP or Unshielded Twisted Pair. UTP is used extensively in telephone networks and in many data communications applications.

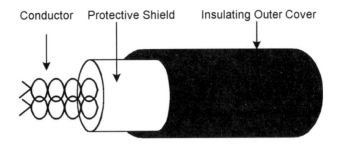

Figure 4-4. Twisted Pair

Shielded twisted pair, like coaxial cable, is able to transfer data faster and over greater distances than unshielded twisted pair because of the additional protection from interference.

Bus Topology

Two topologies are important in LANs: bus and ring. The star topology, as you will see, plays a part in combination with a ring. A bus consists of a single pair of wires (twisted pair) or a wire and a shield (coaxial) which electrically constitute a single circuit (although the bus might be made up of many individual pieces of wire).

At either end of the bus is a terminator, which is essentially a resistor (See Figure 4-5). The resistor is connected between two conductors to provide a load for the circuits that attach to the bus in each node. You might hear of the problem of a network not being properly terminated. This means that the terminator is missing or has the wrong resistance.

The place where a node is connected to a bus is often called a "tap." The bus cable can be tapped: either by cutting it and attaching connectors to the new ends, which is then attached to

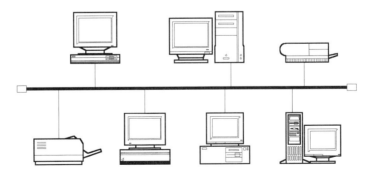

Figure 4-5. Bus Topology

a "T" connector, or by drilling a hole in the cable and screwing a connector in through the insulation until it makes contact with the interior wire of the cable. In any event, the connection is passive. Once the tap is in place, equipment can be installed and removed from the connection without necessarily affecting the network. Most important, if a node fails or stops operating, the rest of the network won't necessarily be affected.

A node on the bus communicates by broadcasting a frame with a header that gives the address of the receiving node. Every node on the bus "hears" every frame. When a node "hears" a message addressed to it, it copies the message off of the bus.

Pure Ring

A ring is made up of a collection of separate point-to-point links, arranged to make a ring. Each node attached to the ring has one input and one output connection, so each node is connected to two links (See Figure 4-6).

Signals received on the input connection are passed through, immediately and without buffering, to the output connection by "repeater" circuitry in each node. Thus, data flow only in one direction on a ring. Each node has the ability to put new bits onto the ring, to send messages, and, if the message is addressed to that node, to copy bits off of the ring as they go by.

In a pure ring, as shown in the diagram, each node is active. If the node fails (for example, if power is removed), then it does not repeat signals from its input. If any node fails to repeat signals, the ring is broken, and data transfer stops until the failing node is restored or removed from the ring.

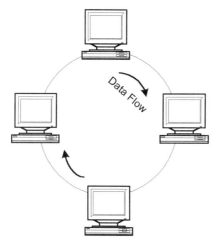

Figure 4-6. Pure Ring Topology

As you might imagine, the ability of a single failing node to bring down an entire ring network is unacceptable. It also turns out that for technical reasons having to do with characteristics of data transmission in a ring, a single ring as we have described it above has a limit of only a few hundred nodes.

Star Ring Topology

The pure ring problems are overcome in the IBM Token Ring (and on the essentially identical IEEE 802.5) by combining ring and star topologies to obtain a more reliable and serviceable configuration (See Figure 4-7). In a star-ring, four wires run from each node to a central ring wiring concentrator, also called a multistation access unit (MAU) or wiring hub.

The MAU detects when a node is not responding and "locks it out" so that the ring can continue to operate when a node fails. This happens automatically when the concentrator senses a node is not responding.

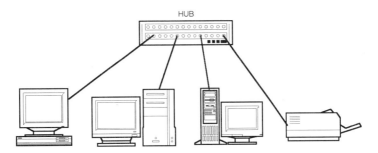

Figure 4-7. Star Ring Topology

The MAU also provides a "bridge" to other rings, sending messages addressed to nodes on other rings across the bridge circuits to those rings, and accepting messages from other rings for its nodes. Rings so joined effectively become a single ring. By bridging wiring concentrators, ring size is effectively unlimited.

A star-ring topology requires more wire to implement than either a pure ring or a bus, but the problems of a pure ring are solved, and a ring topology can have advantages over a bus topology, as you will see later in this chapter.

LAN PROTOCOLS

So far in this section you have seen how nodes can be connected to create a LAN and how signals are sent between two nodes. In this section and the next, you will learn how data, rather than just signals, are transferred across the links.

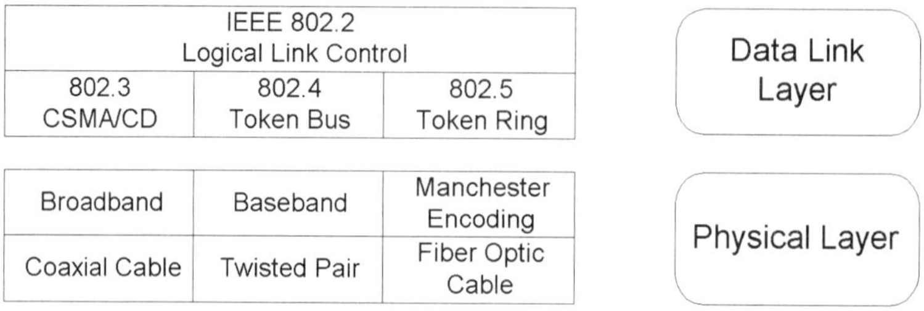

IEEE 802.2 Logical Link Control		
802.3 CSMA/CD	802.4 Token Bus	802.5 Token Ring

Data Link Layer

Broadband	Baseband	Manchester Encoding
Coaxial Cable	Twisted Pair	Fiber Optic Cable

Physical Layer

Figure 4-8. IEEE LAN Standards

The IEEE 802 standards committee produced the 802 standard suite, which has been adopted by the ISO as OSI standard 8802 (See Figure 4-8). The 802 suite includes:

- 802.1 which defines the interface primitives
- 802.2 which defines the upper part of the Data Link Layer and the Logical Link Control (LLC) protocol
- 802.3 which defines the CSMA/CD LAN standard
- 802.4 which defines the Token Bus standard
- 802.5 which defines the Token Ring standard

You will recall from Chapter 2 that the OSI's data link protocol for point-to-point communication is HDLC-High Level Data Link Control. The originators of the 802 standards saw the need to create a sub-level above the bus and ring protocols to provide the same services to the Network Layer for LANs that HDLC provides for point-to-point links. To satisfy that need, they created LLC. Its job is to accept packets from the Network Layer, divide them as necessary into frames, and then hand them off through the MAC layer to the appropriate bus or ring process for transmission, while handshaking with its peer process to ensure error-free delivery.

LAN PROTOCOL ISSUES

The process of communicating on a LAN falls into two parts: Media Access Control and the Data Communication Protocol.

Medium Access Control

A procedure by which a node gains control of the medium must be in place. With a bus architecture, when a node transmits, every other node on the network (or at least the subnet) receives the transmission. When two nodes attempt to transmit simultaneously, a collision, or garbled transmission, results. Since it is possible that two nodes might wish to transmit at the same time, some means must be available to either prevent collisions or handle them properly when they occur.

With the ring architecture, only one node transmits on a given link on the ring, so the problem of garbled transmissions does not exist. But remember that each node must immediately repeat data received on its input connection. How then does it get to transmit its own data when it needs to, if the ring is busy? That is the problem of medium access control in a ring architecture.

Data Communications Protocol

A protocol must also exist for transmitting data once the node has gained control of the medium and is able to transmit.

APPROACHES TO MEDIA ACCESS CONTROL

This section deals with the lower sublayer of the Data Link Layer. This lower sublayer, also referred to as the Media Access Control (MAC) layer, is the framing and access control layer for the Token Ring protocol.

The problem with media access is what to do with collisions and the garbling or loss of messages that results when collisions happen. There are two ways to approach the problem: let the collisions happen and then worry about it; or devise a protocol that allocates or schedules time for each node that wants to transmit. In fact, these two general approaches represent the two classes of protocols we are studying in this chapter, Ethernet and Token Ring.

IEEE 802.3 Ethernet allows collisions to occur but ensures that they do not happen too often and makes sure that, when they do occur, no data are lost or damaged.

Token passing (as implemented in both Token Ring and Token Bus) does not allow collisions to occur and ensures that each node is given the opportunity to access the LAN.

ETHERNET (802.3)

Ethernet technology, originally developed in the 1970s by Xerox Corporation in conjunction with Intel and Digital Equipment Corporation, is now the primary medium for local area networks. Ethernet has 10 Mbps throughput and uses the CSMA/CD method to access the physical media.

Standard Ethernet equipment is now very inexpensive. Network interface cards (NICs), hubs, and cabling have become commodity items. It is estimated that over two-thirds of all current Local Area Networks are Ethernet.

Ethernet Media Access

The inventors of Ethernet chose a technique called Carrier Sense Multiple Access with Collision Detection, or CSMA/CD, as their technique for controlling access to the bus.

An Ethernet node must have the ability to listen to the bus and transmit at the same time. To transmit a message, the following logic is applied (See Figure 4-9).

1. Listen to the bus to see whether any other node is transmitting. This is done by sensing the carrier signal. If transmission is detected, continue listening until the channel is idle.
2. When no signal is detected, start transmitting the message.
3. While transmitting, also listen to the bus. Compare the received message to what was transmitted. As long as they are the same, continue transmitting.
4. If what is received is not what was transmitted, a collision is assumed. Stop transmitting.
5. Transmit Jam sequence to warn all other stations that a collision has been detected.
6. Wait a random time and then start over with step one.

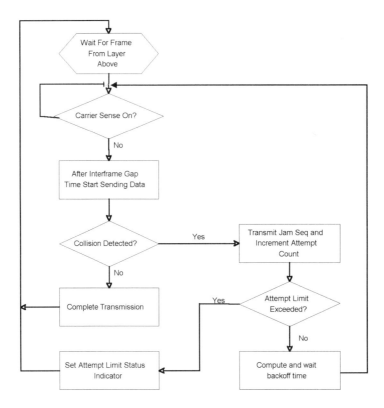

Figure 4-9. Ethernet Send Algorithm.

An important aspect of the CSMA/CD algorithm is the random-length interval (a few milliseconds) that the node waits before trying to retransmit when a collision occurs. Because the probability of both nodes waiting the same length of time is very small, a second collision is unlikely to occur, ensuring that nodes don't just keep "butting headers," with none able to access the bus.

On a heavily used network, collisions will occur, and some time will be wasted each time this happens. This is one reason why performance of the network decreases as the load increases.

Note that the node stops transmitting as soon as a collision is detected—that is the significance of the "CD" (collision detection) in CSMA/CD. Other similar protocols don't require the node to stop transmitting until the end of its message, so a great deal of time is lost when any of the colliding nodes is transmitting a long message.

A potential problem with CSMA/CD is the time it takes for an electrical signal to travel from one point on the bus to another. In a large network, if two nodes, A and B, are at opposite ends of the bus, and both start transmitting at the exact same moment, node A might transmit the last bit of its message before the first bit has had time to get to B, and vice versa. Neither would detect a collision, yet the signal would be garbled for some or all of the nodes in between. One way of solving this problem is to require that messages be at least long enough to prevent the situation. This is how Ethernet handles it.

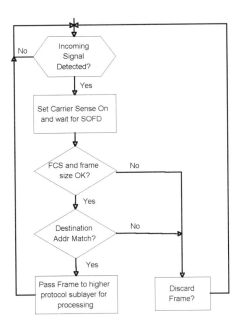

Figure 4-10. Ethernet Receive Algorithm.

The Ethernet Receive Sequence diagram (Figure 4-10) shows the sequence for receiving a frame. To receive a frame on Ethernet, the node (NIC) first detects the signal. Once detected, carrier sense is turned on to prevent transmission of another frame. The frame is received and first checked for errors and then for minimum length. Once completed the destination address is examined to see if it matches this node. When a match occurs, the frame contents are passed up to the Network Layer.

Ethernet Frame Format

The Ethernet Frame format is shown in Figure 4-11. Ethernet messages contain these fields:

- Preamble-seven bytes of "10101010" to synchronize with all other stations on the LAN
- Frame Delimiter (Also called the Start of Frame Delimiter) 10101011
- Destination Address—48 bits (Destination NIC)
- Source Address—48 bits (Sending NIC)
- Length of Data
- Data—0 to 1,500 bytes—48 bits
- Padding-if length of data < 50 bytes, pad message to 50 bytes
- Checksum (FCS)

The padding of the message has important implications, as you will see later in the chapter. Padding is done to ensure that the message is long enough that the preamble gets to the farthest node before the end of the message is transmitted, as explained above. The illustration we have used here is a worst-case example. It is quite common because some very heavily used applications, such as the UNIX Visual Editor (vi), are character-oriented, so each character is transmitted when the key is pressed. (Applications like FTP transmit blocks most of the time, so padding is not a problem).

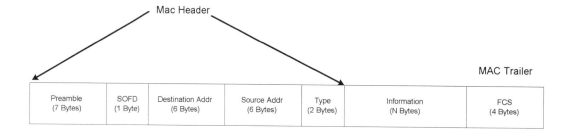

Figure 4-11. Ethernet MAC Frame.

Ethernet messages can be addressed to a group of nodes by setting one of the control bits in the destination address. This allows "multicasting" of messages. The Ethernet address field can be a global address (indicated by the other control bit). Any Ethernet node in the world can be addressed in this way, although the Network Layer must take care of messages to destinations not on the LAN.

Ethernet Configuration Example

An example of an Ethernet configuration is shown in Figure 4-12. In this case, several nodes are physically attached to a LAN cable or bus, and transmit frames to and from other nodes connected to the same bus. The three nodes and the server share the available 10 Mbps bandwidth. This diagram also shows an Ethernet hub connecting three workstations and a server. This is a star configuration and is found in Ethernet networks using 10Base-T technology. All of the nodes attached to the hub share the same 10 Mbps bandwidth.

Ethernet at 100 Mbps

It is amazing to think that a technology in today's world could last two or three decades and still have the widespread acceptance such as found with Ethernet. For the typical user, the 10 Mbps

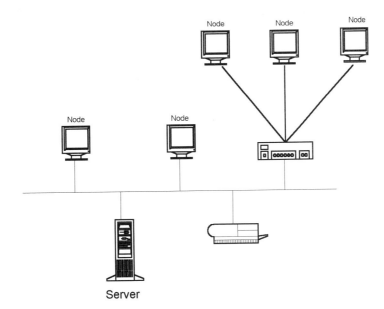

Figure 4-12. Sample Ethernet Configuration.

rate of Ethernet continues to provide plenty of bandwidth, even when shared with other LAN users. However, as the number of client/server applications grows, and as the content of data changes, 10 Mbps no longer remains acceptable in many networks. The need to transmit voice, data, and graphic images simultaneously and the increasing sophistication of the typical user has driven the need for more bandwidth and greater responsiveness by the network.

There are several methods used to increase bandwidth on Ethernet based networks. The use of bridges, routers, and switches to increase performance are covered in the internetworking chapter. Essentially, these devices isolate local traffic on multiple Ethernet subnets. Ethernet performance can also be enhanced by using newer technologies which provide Ethernet at 100Mbps. There are two current 100 Mbps standards referred to as 100Base-T or Fast Ethernet, and 100VG-AnyLAN.

100Base-T Ethernet

The IEEE 802.3u committee is currently responsible for the development of the 100Base-T standard. As the committee name implies, this standard is almost identical to the IEEE 802.3 Ethernet standard in functionality. The similarities between standard Ethernet and Fast Ethernet are listed below:

- The CSMA/CD Access Method is used
- It runs over unshielded twisted pair cabling
- It maintains the same physical architecture

100Base-T using twisted pair cabling can extend to distances of 100 meters. 10 Mbps Ethernet can extend 500 meters. The reason for the decreased length of 100BASE-T is to provide collision detection at the increased speed. The use of fiber optic cabling can extend the distance to 325 meters. An example of 100Base-T network is shown in Figure 4-13. This configuration is similar to a star configuration used in 10 Mbps Ethernet LANs. The devices connected to each of the 100Base-T hubs share the 100 Mbps bandwidth. It is also possible to combine Ethernet LANs using both 10 Mbps technology and 100 Mbps technology. This is shown in the second 100Base-T diagram.

100Base-TX is the specification that describes how to run 100 Mbps Fast Ethernet over Category 5 Unshielded Twisted Pair. Category 5 UTP is the most popular type of cabling used in LANs today.

The MAC Frame of 100Base-T and 10 Mbps Ethernet are identical. Therefore, equipment used to bridge 10 Mbps Ethernet with Fast Ethernet must only be concerned with the data rate, and not the content or format of the frame. Other key points to mention are:

- The 100Base-T standard requires replacement of existing equipment such as NIC cards and sometimes cabling.
- The maximum UTP cable length is 100 meters.

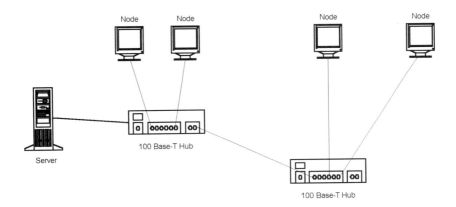

Figure 4-13. 100Base-T Configuration

- Category 5 cabling must be used.
- The 100Base-T standard can extend distance capabilities by using fiber optic cable technology.
- The maximum number of repeater hops is 2.

100VG-AnyLAN

100VG-AnyLAN is another 100 Mbps Ethernet technology standard. It directly competes with 100Base-T Ethernet. The access method used by this standard is different than 10 Mbps Ethernet and Fast Ethernet (CSMA/CD). The MAC Frame, however, stays the same. The new access method is called "demand priority." Refer to Figure 4-14 for the following discussion of demand priority.

The Demand Priority Access Method (DPAM) consists of two levels of requests that can be made from network users. Normal priority requests are for standard data transmissions and high-priority requests are for time-sensitive data such as multimedia. Each frame sent out over the LAN is assigned a priority by the application layer.

100VG-AnyLAN topologies consists of three primary components:

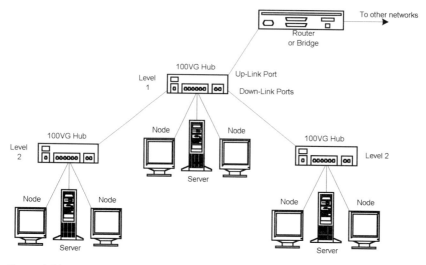

Figure 4-14. 100VG-AnyLAN

- Point-to-Point Links
- Network Nodes
- Hubs

The hub is the crucial component in a 100VG-AnyLAN network. It is the responsibility of the hub to manage the network nodes. Each hub polls each node attached to it to see if there is data to send across the network and determine the priorities of the tasks. Hubs can be root hubs (Level 1) or lower-level hubs. Lower-level hubs are viewed by the root hubs as nodes.

Several factors should be considered when implementing 100VG-AnyLAN networks:

- There can be a maximum of three levels in any 100VG-AnyLAN network.
- Using Category 5 cabling, 100VG-AnyLAN segments can extend up to 150 meters (100 for 100Base-T).
- A maximum of 1024 nodes are allowed under one root hub.
- Bridges and/or routers may be used to connect different networking technologies.

TOKEN RING (802.5)

The IEEE 802.4 and 802.5 standards and the IBM Token Ring protocol (with which IEEE 802.5 is compatible) use token passing, an altogether different approach than CSMA/CD, for medium access control. We will consider first the case of token passing in a ring (802.5) and will treat Token Bus (802.4) briefly, later in this section.

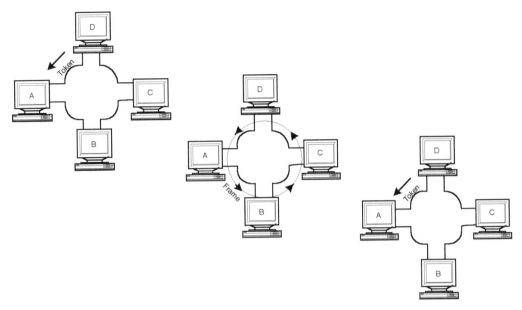

Figure 4-15. Token Ring Media Access

Token Ring Media Access Control

Recall that in ring networks, each node contains a repeater that receives bits from one of the two links and transmits them on the other. It receives messages simply by copying bits as they go by. The medium access control question with a ring LAN is—When can the node insert bits onto the ring? The answer lies in the token passing protocol (See Figure 4-15 and the description below).

The idea of token passing consists of several elements:

- On an inactive LAN, a three-byte token circulates endlessly.
- The token is like a frame, except that bit 4 in the second byte indicates that it is a token.
- Three priority bits indicate whether the token can be grabbed by the station. When the priority of the token is higher than the frame to be transmitted, the token is passed on.
- A node transmits its message by inserting it after the token when the token is free.
- Each node retransmits the message that follows a busy token.
- Only one token, and at most one message, can circulate on the ring at one time (for 802.4 or 802.5; not true for FDDI, as you will see later).

Token Ring Protocol

When a node wants to transmit a frame, it grabs the token and appends the remaining fields onto the frame. The node must stop transmitting after ten milliseconds and wait for the token to come around again if it has more data to send (Refer to Figure 4-15).

The token will circulate around the ring, followed by the message, passing through the destination node, until it returns to the source node.

As it passes through the destination node, the node copies the data into its own memory (The A and C bits in the frame status byte are used to signal the sending node that the node is alive and the frame has been copied).

When the source node receives the busy token, it retransmits it as a free token. As it receives the data following the token, it does not retransmit it, thus removing the message from the ring. Figure 4-16 illustrates the general process required for a node to receive a frame.

Early Token Release

The Early Token Release (ETR) feature is a method of releasing the control token on a Token Ring LAN before the sending station has received the transmitted data frame. This is done on 16Kbps Token Rings.

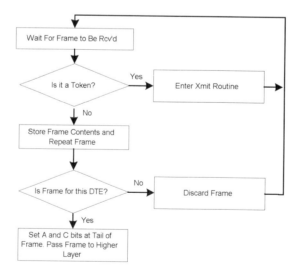

Figure 4-16. Token Ring Receive Algorithm

Monitor Function

An advantage of token passing is built-in acknowledgment. When the node that sent the message gets it back, it knows that the message has traversed the ring and should have been received by the addressee. It can check the C-bit to see if the frame has been copied by the receiving node. Note that for IEEE 802.5, only one token and only one message can be on the ring at any time, although different parts of the token and message are being simultaneously received and transmitted by the nodes.

A disadvantage of token passing is its susceptibility to certain types of errors. When a node fails, it can leave a busy token endlessly circulating, which would idle the whole network. It can also fail to transmit a free token, with the same effect. Because of these and other potential problems, each Token Ring LAN must contain a node which, in addition to its normal functions, performs a monitor function. This node watches traffic on the ring to detect failure modes and take corrective action, such as inserting a free token when necessary. Of course, having the monitor function introduces the problem of what to do when that node fails. . . and so on.

Token Ring Configurations

This section illustrates two Token Ring configurations found in networks today. Figure 4-17 illustrates how several ring segments are connected together via Multistation Access Units (MAUs). Note that larger rings can be built by connecting the Ring Out (RO) connection of one MAU to the Ring In (RI) connection of another MAU.

Figure 4-18 illustrates three traditional Token Ring configurations used to access an IBM host. Token Ring nodes can access host resources via Front End Processors, Cluster Controllers, or through LAN Gateway devices such as a IBM 3172. These are explained in the System Architectures chapter of this book.

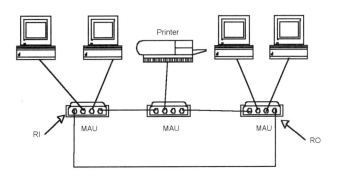

Figure 4-17. Token Ring Configuration using an MAU.

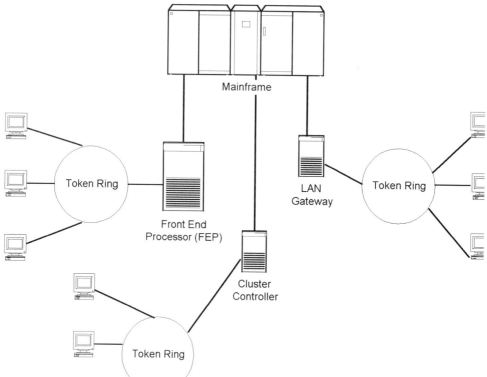

Figure 4-18. Token Ring and IBM Host Connectivity

Token Ring Frame Format

Figure 4-19 shows the Token Ring Frame Format. The top portion of this diagram shows the format when data from an upper layer is being encapsulated by the Token Ring MAC header and trailer. The bottom portion of the diagram shows the format of a token. When nodes on a Token Ring segment have no data to send, the token circulates around the ring. When a node has data to send, it first receives and repeats the Starting Delimiter. It then checks the token priority in the Access Control Field to make sure a higher priority task on the ring is not waiting for the token. If the priority is ok, it appends the remaining header information, frame data, and trailer. The Token Ring frame contains the fields shown below:

- SD: The starting delimiter byte indicates the start of the Token Ring frame.
- AC: The access control field contains the priority bits (P and R), monitor bit (M), and token bit (T). The format of this byte is PPPTMRRR.
- FC: The frame control byte format is FFZZZZZZ. It indicates whether the frame contains

information from layers above (Fs) or is link layer control information. Control information is also contained in this byte (Zs).

- DA: The destination address contains the address of the node or nodes to receive the data. It can be 16 or 48 bits in length.
- SA: The source address contains the address of the sending station. It can be 16 or 48 bits in length.
- INFO: The information field contains either data from an upper layer or additional control information beyond what is contained in the frame control field.
- FCS: The Frame Check Sequence is used for error control. It covers FC through FCS.
- ED: The ending delimiter.
- FS: The frame status byte contains the A (Alive) and C (Copied) bits. If the A bit is set, it indicates the receiving station(s) is alive. If the C bit is set, it indicates that the receiving station(s) has copied the frame.

Token Ring Advantages

1. Simple engineering because it is point-to-point digital—no analog
2. Standard twisted pair medium is cheap and easy to install
3. Easy detection and correction of cable failures
4. Deterministic and able to prioritize traffic
5. No padding of data required in frame, so frames are short
6. Excellent performance under conditions of heavy load
7. Since rings can be bridged by their wiring concentrator into what is effectively one ring, ring size has no practical limit

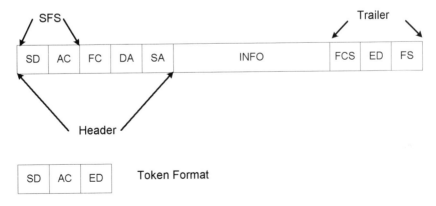

Figure 4-19. Token Ring Frame Format

Token Ring Disadvantages:

1. Necessity of having a monitor function
2. Under conditions of low load, substantial delay waiting for token to come around, even though the network is idle
3. Can require significantly more wire to be run than a bus architecture

TOKEN BUS

You will recall that on a CSMA/CD LAN, nodes vie for the ability to transmit messages on a first-come-first-served basis, where ties are broken by having both nodes wait a random time interval. A problem with this approach is that under conditions of maximum network activity, a node can become "unlucky" and not be able to transmit for a long period of time. CSMA/CD provides no way to guarantee that the wait time won't exceed a given maximum; it is a matter of probability and network load.

The Token Bus protocol was designed to solve this media access control problem. The basic idea of Token Bus is to treat the bus as if it were a "logical ring." The following diagram shows how this is accomplished (See Figure 4-20):

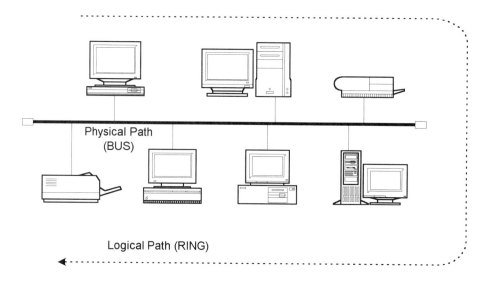

Figure 4-20. Token Bus

1. Each node knows the network address of its successor in the logical ring.
2. A token is passed from node to node, each node passing the token to its successor on the ring. As with Token Ring, a token looks like a message, except for a header bit which says "I am a token."
3. There is only one token, and one node which has the token, called the "token holder" which can transmit. All other nodes can only receive.
4. When a node becomes the token holder by receiving the token, it has a period of time available to it during which it can transmit if it wishes.
5. If the token holder does not need to transmit, it just passes the token to its successor.
6. If the token holder needs to transmit, it can do so until its time interval has expired; then it must stop transmitting and send the token onward. This implies a maximum message length, with segmenting of long messages.

As you can see, each node is guaranteed the opportunity to transmit within some defined maximum interval. This is not the case with Ethernet bus. This guaranteed access to the network is important for the environment which the Token Bus protocol is aimed—at the factory floor. However, this predictability has an associated cost, since a significant amount of time that could be used to transmit data messages on a CSMA/CD bus is wasted passing tokens on a Token Bus.

Token Bus Advantages

1. Uses cable TV cables and parts readily and cheaply available
2. Deterministic and able to prioritize traffic
3. Short minimum frames
4. Excellent performance under conditions of high load
5. Broadband can support multiple channels (video, voice)

Token Bus Disadvantages

1. Complex protocol and engineering of equipment
2. Expensive—requires modems and repeaters
3. Since node must wait for the token to come around before transmitting, messages are delayed waiting for the token even when network is idle

FDDI

In considering the use of fiber optics for the transmission medium in a LAN, one might think that the fiber optic medium might simply be substituted for copper to provide greater speed of transmission. In fact, this can and has been done. But the Fiber Distributed Data Interface (FDDI) allows for more than just a faster medium.

FDDI is a ring protocol, based on the IEEE 802.5 Token Ring standard. An FDDI ring can be combined with 802.5 rings or 802.3 buses to provide a high throughput "backbone" for a network (we will discuss combining rings more in the next section).

Although FDDI is similar to 802.5, three important differences make the technology and protocols of FDDI more complicated (See Figure 4-21):

1. An 802.5 ring can have only one physical link between two nodes; FDDI has two links, with data circulating in opposite directions.
2. An 802.5 ring can have only one token, and at most, one message circulating at one time; FDDI can have many on each ring.
3. A portion of an FDDI ring data transmission capacity can be given up for "synchronous" traffic, essentially layering one or more high-speed communications channels onto the ring.

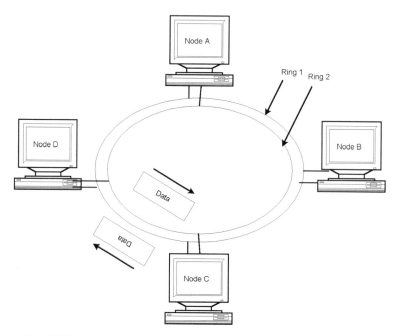

Figure 4-21. FDDI

FDDI also supports multiple frames. Consider the case of a single message transmitted from a node around a ring back to the node. If transmission were truly instantaneous, the node would see the first bit of the message on its input connection the moment it transmitted it on its output connection. But it takes time for a signal to make its way around the ring—time for the signal to get from one end of each segment of the ring to the other and time for the repeater to copy each bit from its input to its output. As many as 1,000 nodes and up to 200 km of fiber might be involved in an FDDI backbone ring. There can be a long delay between the time when a node transmits the final bit of a message and when the first bit of the token is received. In an 802.5 ring of that size, this time is wasted, because the token must come all the way around and be freed by the originating node before any other node can transmit.

In an FDDI ring, when a node finishes transmitting a message, and it is not receiving anything, it immediately transmits a free token, essentially creating a new token where there was none before. That token can then be used by the next node on the ring that needs to transmit, and that transmission will occur simultaneously with the transmission of data for other busy tokens on the ring.

Dual Channels

The FDDI standard specifies dual channels, running in opposite directions. The purpose of this requirement is to provide much greater reliability and recoverability than would be possible with a single link. This is obviously a highly desirable goal for a standard aimed at network back-bones. The dual channel architecture is implemented as follows:

A node can communicate on one or two of the channels (the standard does not require communication on both channels, so lower-cost, single-channel machines can be built). If the node communicates on both channels and one channel fails, the node often will still be able to communicate on the other. The two types of FDDI stations are referred to as Dual-Access Station (DAS) and Single-Access Station (SAS).

Each node on the net, whether it communicates on one channel or on both, must still connect to both and provide shunt circuits to handle recovery.

When a break in both channels on the ring occurs, either because of a node failure or because a link is damaged, the nodes to either side of the break or failing node connect the two channels together, effectively turning them into one longer ring. This allows operation to continue until the problem is resolved.

Synchronous Traffic

If an FDDI ring serves as the backbone for a large network, for example, in a large engineering facility such as one might find in an aerospace firm, then network traffic might be far less at night than during the day. At night, a significant portion of the FDDI ring capacity goes unused. For reasons such as this, the FDDI standard allows a portion of the 100 Mbits/second bandwidth of the ring to be given up to emulate "T1" or "T3" communications channels. Tl and T3—El and E3 in Europe, are high-speed digital communications channels that can be used for either voice or data.

LOGICAL LINK CONTROL PROTOCOL

As discussed earlier in the chapter, the Logical Link Control (LLC) is the top sublayer of the 802 data link layer. The LLC layer is independent of the MAC access medium and protocols as well as the network layer protocols above.One function of the LLC layer is to direct frames from the LAN to the appropriate network layer protocol. In other words, the LLC layer can be used with Token Ring or Ethernet sublayers and with multiple network layer protocols (See Figure 4-22).

An example of an LLC Protocol Data Unit (PDU) format is shown in Figure 4-23. It contains the following:

- DSAP or Destination Service Access Point
- SSAP or Source Service Access Point
- Control Field
- Information Field

Note that this is an Information Frame. There are two other possible frame types, supervisory and unnumbered. The LLC protocol is based on the HDLC control, which is discussed later in the chapter on telecommunications. The LLC protocol is made up of three forms of service:

1. Unacknowledged connectionless service.
2. Connection-mode service.
3. Acknowledged connectionless service.

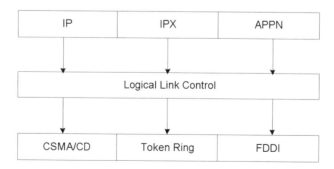

Figure 4-22. Logical Link Control Function

DSAP	SSAP	Control	Information

Figure 4-23. LLC Frame Format

LLC provides two types of flow control:

1. Stop and Wait. Acknowledgment is sent back after each frame is received. Sender must wait until acknowledgment is received before sending the next data unit.

2. Sliding Window. Allows multiple PDUs (Protocol Data Units) to be sent at one time. If a receiving station allocates seven input buffers it can accept seven PDUs. To keep track of the acknowledged PDUs each is labeled with a sequence number from 0 to 7. The receiving station acknowledges receipt of a PDU by sending the number of the next PDU expected. If two stations are sending and receiving data, two windows must be maintained, one for transmitting and one for receiving. Acknowledgments and data can be sent together; this is known as "piggy-backing".

Figure 4-24 illustrates the concept of the sliding window. In this diagram two nodes are exchanging PDUs. Node X is sending Information to Node Y and vice versa. There are four items depicted in the diagram as being sent from Node X to Node Y. They are the DSAP (X or Y), the PDU type (Information, Unnumbered or Supervisory), the send count and the receive count. The sequence goes like this:

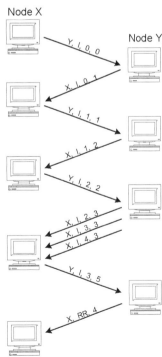

Figure 4-24. LLC Sliding Windows

1. Node X sends PDU 0 and expects PDU 0 from Y
2. Node Y sends PDU 0 and expects PDU 1 from X
3. Node X sends PDU 1 and expects PDU 1 from X
4. Node Y sends PDU 1 and expects PDU 2 from X
5. Node X sends PDU 2 and expects PDU 2 from Y
6. Node Y sends 3 PDUs (2,3 and 4) to Node X and expects 3 from X

7. Node X sends PDU 3 and expects 5 from Node Y
8. Node Y sends an unnumbered PDU (RR, no information to send) to X and expects 4 from Node X

Note that in step 7 Node X acknowledged 3 PDUs while also sending information. This illustrates the sliding window concept as well as piggybacking.

SAPs are locations where the sending and receiving nodes can store information for each other. Each SAP has a one byte address associated with it. The same LLC layer may have to provide a communications path for multiple network layer protocols such as IPX or IP.

Summary of LAN Protocols

As you can see, the choice of a standard will depend on the particular installation. Factors like the difficulty of installing wiring, wiring already in place, requirements for throughput requirements for prioritizing traffic or having deterministic network access delays, and so on, will all play a role in the decision. The architecture will be chosen based upon the "fit" to the environment. The only clear and unequivocal difference among the architectures is that Ethernet degrades severely under heavy load (effective throughput drops off due to collisions), but Token Bus and Token Ring handle heavy loads gracefully, continuing to provide highly effective throughput. Approximately two-thirds of installed networks use Ethernet as the LAN technology.

WIRELESS LANS

Over the past few years a new trend in Local Area Networking has developed. This new technology is typically referred to as Wireless Local Area Networks (WLANs), but is also referred to as Local Area Wireless Networks (LAWNs). WLANs are identical to other LANs with the exception that data is transmitted and received without the need for wires (See the Simple WLAN diagram). A great deal of work is being done by groups such as the IEEE 802.11 committee to develop standards and protocols which will enhance the growth and development of this new technology. Since more and more people have accepted the convenience of cellular phones, it is believed that people will also see the benefit of linking portable computers through such networks.

There are three primary techniques used for sending and receiving data using wireless technology in Local Area Networks. The primary difference between these techniques is the frequency that is used to transmit the signal between stations. Each of the technologies could be employed; however, there are advantages and disadvantages of each. These three techniques are described in the following sections.

Spread-Spectrum Radio

Spread-spectrum radio frequencies are not regulated by the FCC and have been designated as the industrial, scientific, and medical (ISM) bands, although they are not limited to these industries. Spread-spectrum was initially developed for use by the military in World War II. It occupies three radio frequency bands: from 900MHz to 928MHz, 2400GHz to 2483.5 GHz, and 5725GHz. There are two primary types of spread-spectrum: direct sequence and frequency hopping. Direct sequence refers to the ability to send a modulated signal across several spectrum frequencies rather than just a single frequency. This provides some level of security since the receiving device must be configured to detect the exact pattern as the sending device is generating. Frequency hopping causes the modulated signal to hop from one frequency to another. The receiving node must be synchronized with the sender and must know the frequency sequence being used.

Infrared Light

Infrared frequencies operate at a much higher frequency than radio frequency (RF) WLANs. Typical IR frequencies are in the 1,000 GHz (TERA) range and above. Infrared can be used as a tightly focused beam of light for point-to-point communication or as a diffused beam of light which can cover an entire room (broadcast approach). One of the primary problems with infrared light is the lack of range, because any obstruction (such as walls) will not allow the signal to reach the receiving station. The positive side of this limitation is security. Infrared is also immune to electronic interference. The use of infrared is not controlled by any agency and no license is required to develop or operate infrared LANs.

Narrow-Band Radio

This technique is similar to radio transmission used by radio stations. The transmitter and receiver operate on the same frequency by "tuning" to the same "station." The signal can go through walls and can carry over long distances. The major problems with narrow-band radio are the tuning that is required to synchronize the stations and signal reflections which can cause interference problems.

Wireless LAN Configurations

Wireless LANs can be configured in many different ways. This overview presents three possibilities. The first possible configuration would be a traditional Ethernet or Token Ring LAN without the cables. In other words, a LAN with no wires. The physical media, such as Ethernet

10Base5 cabling, is replaced with wireless transmitters and receivers.

Figure 4-25 shows one wireless LAN configuration possibility. In this diagram, two "hard-wired" LANs are combined or interconnected through the use of wireless technology. Two Ethernet LANs are shown connected by a wireless link. The distance between the two wireless antennas in this configuration varies depending on the technology used.

Figure 4-26 is another example of a wireless configuration. This configuration would be used by remote PC users who would like to access a LAN remotely through a laptop PC without having to use a phone jack. These PCs would be equipped with wireless LAN adapters and users could gain quick access to remote LANs via their portable PC.

Wireless Technology Advantages

1. Reduced installation time and labor
2. Elimination of cabling and cabling closets
3. High reliability
4. Convenience

Wireless Technology Disadvantages

1. Limited distances
2. Limited transmission speeds
3. High cost per unit installation
4. New and "unproven" technology

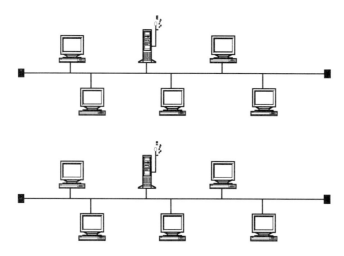

Figure 4-25. Server Connected Wireless LAN

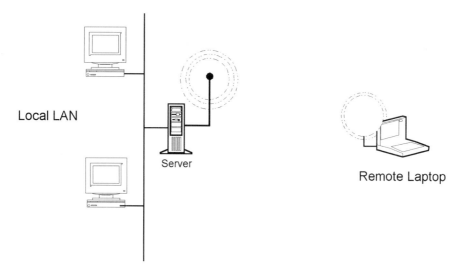

Local LAN

Server

Remote Laptop

Figure 4-26. Wireless connectivity of remote laptop

SUMMARY

When the first "personal" computers were networked at Xerox Parc in the 1970s, the available networking technology was found wanting in many respects, and Ethernet was born. As workstations proliferated in the early 1980s, so did Ethernet, and it spread to the networked PC environment. Somewhat later, IBM produced a competing standard, Token Ring, based on technology developed for networked controllers. In the mid-1980s, IEEE adopted both protocols as standards as part of its 802 protocol suite, adding the Token Bus architecture needed for its deterministic nature on the factory floor. In 1990, FDDI was adopted by ANSI as the high-speed standard for backbone networks. Refer to the LAN Table for the following discussion.

LANs can be based on coaxial, shielded or unshielded twisted pair, or fiber optic. Ethernet most often uses coaxial but can use twisted pair; Token Ring uses shielded twisted pair; and FDDI uses fiber optic cable.

A bus topology connects every node to the same electrically common cable. Ring technology is a series of point-to-point channels, arranged logically into a ring, with each node receiving on one channel and transmitting on the other. Rings are actually implemented as star rings to get around the problem that occurs in pure rings. If a node fails, the network goes down. FDDI is a ring technology with dual rings, with the data flowing in opposing directions on each ring. When a node or channel fails, the dual rings can be combined to form one big ring, with the failing component shunted out of the ring until the problem is resolved.

In an Ethernet LAN, the CSMA/CD protocol allows nodes that are trying to transmit at the same time to detect the collision, halt transmission, and start over after a random interval. They aren't likely to collide twice in a row, except when the network load is heavy. Because of collisions, Ethernet does not perform well when the network is heavily loaded.

Token Ring manages access to the ring through the passing of tokens. To transmit a frame, a node must wait until a free token passes. When it sees a free token, it can change it to a busy token and insert a frame onto the ring. The message is copied by the addressee and removed from the ring when it comes back to the originator.

FDDI, aimed at backbone networks, improves on 802.5 Token Ring by using dual channels. Every FDDI node must contain circuits that can shunt the two rings together, so that when a node or channel fails, the two rings become one and operation continues, at a somewhat slower pace perhaps, until the failure is corrected. FDDI also allows more than one message to be circulating on the ring at one time.

The principal limits to LAN performance and capacity are loss of signal strength over distance and the delay inherent in the propagation of signals over a wire. Bus protocols like Ethernet are not capable of much higher data rates than they currently provide, because propagation delays stay the same as data rates increase and cause unacceptable loss of efficiency.

Ethernet provides throughput of 10 Mbits/second over a distance of up to 2.5 km. Token Bus can operate at 1, 5, or 10 Mbps at up to 7.6 km. Token Ring can operate at 1, 4, or 16 Mbps, depending on medium, with up to 260 nodes on a ring. FDDI operates at 100 Mbps on each of the two rings, and the rings can span a distance of 2000 km with a distance of 2 km between nodes.

Ethernet is the lowest cost, simplest to install, and most popular of the standards, but network capacity is wasted for short messages, and performance degrades under heavy load. Token Ring performs predictably, and throughput is sustained even under a heavy load. In a star-ring configuration, failures are easy to locate compared to Ethernet. Token Bus provides the low cost of installation of Ethernet with the deterministic performance of Token Ring for the factory floor.

LAN Performance, Capacity Selection, and Use

Two primary factors limit the performance of LAN technology; these are summarized below:

1. Signal Attenuation. All transmission media have some resistance to the flow of electromagnetic force. Over a great enough distance, the signal will be attenuated (reduced) to the point that an unacceptable number of errors is introduced. Fiber optic transmission is far less susceptible to attenuation than electrical transmission. Broadband is less susceptible than baseband.

2. Signal Propagation Delay. Even at 186,000 miles per second, the time that it takes for a signal to propagate from one end of the network to the other is an important limiting factor.

These and other factors limit LANs in two ways: the speed at which they can transmit data reliably, and the distances over which they can do so.

Throughput improvement for IEEE 802.3 (Ethernet) in the future is especially limited by propagation delay. As mentioned earlier the current data rate of 10 Mbits/second requires a minimum frame size of 50 bytes in order for all collisions to be detected. This minimum would have to increase in proportion to any increase in transmission speed. A minimum frame size of 400 bytes at 100 Mbits/second would clearly be an unacceptable waste of the channel bandwidth.

ADDITIONAL INFORMATION ON THE CD-ROM

- LAN Architecture Test Questions
- Additional Information on terms and concepts covered in this chapter
- Traces of Ethernet LAN Traffic
- Traces of Token Ring LAN Traffic
- Additional Ethernet Configurations
- Additional Token Ring Configurations
- Additional Cable Descriptions
- Cabling Specifications and Categories

5

NOS ARCHITECTURES

INTRODUCTION TO NOS ARCHITECTURES

All of the preceding chapters have been designed to give you an overview of the networking standards that have merged as a result of the response of large companies and other organizations, such as government agencies, to users' demands that their departmental systems, workstations, and PCs be integrated into the larger organizational environment. TCP/IP was the result of users in research institutions worldwide creating a mechanism for communication among their peers. Digital's own internally developed DNA was its response to its own users' demands.

At the same time that major vendors were working to protect their minicomputer and mainframe installation base by offering networking solutions for PCs, a new, unknown group of vendors responded to the opportunity that existed in the "personal" computer market—an opportunity recognizable wherever users talked about "islands of technology" and "sneakernet." Today, most professionals don't have to ask who Novell, Microsoft, and Banyan are. Their network solutions, along with solutions provided by Apple, IBM, and others, have revolutionized the role that PCs and Macintoshes play throughout the corporation.

Although LANs play an increasingly important role in corporate environments, the bulk of the PC user base is still "networked" simply for the sharing of peripherals. This represents an enormous opportunity for PC LAN vendors, because the migration of LANs into the network computing environment of corporations will continue to drive the networking of PCs, Macintoshes, and workstations. At the same time, in order to protect their installed base, systems vendors like IBM and DEC have entered the LAN marketplace, making the race to network users even more competitive and interesting.

In this chapter, you'll be guided through a brief description of the most important PC LAN solutions. We will examine how PC LAN Network Operating Systems (NOS's) relate to the OSI Model, and look at the key elements of each NOS vendor's products.

CHAPTER OBJECTIVES

At the completion of this chapter, you should be able to:

- Name the advantages of the client/server model when used in PC LAN environments

- Look at Remote Procedure Call tools used by developers of network applications

- Identify the generic mapping of PC LAN Network Operating Systems to the OSI model

- Identify primary products and product characteristics of the leading NOS vendors

- Identify major trends in the PC LAN marketplace

THE CLIENT / SERVER MODEL

In Chapter 1 we discussed three architectures which applications can use to distribute functions across a network: client/server, peer-to-peer, and master/slave. Let's look more closely at the client/server model (See Figure 5-1). Most of the PC LANs we will consider use the client/server approach to resource sharing. The client/server model essentially divides a task into two parts and executes each on a different system on the network. For example, the task of creating a report can be divided into the application portion, where the report is created (client process), and the server portion, where the report is printed. The client computer is often referred to as the "front-end" system and the server as the "back-end" system.

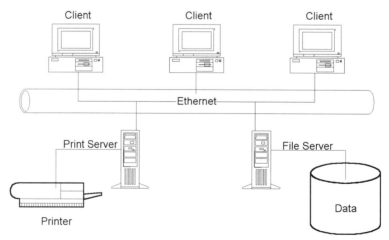

Figure 5-1. Client/Server Diagram

The client and server processes interact with each other by transmitting request/reply pairs. The client process initiates an interaction by issuing a request to the server. The server process responds with a reply satisfying the request. If a request cannot be satisfied, the server process provides an error message.

A client/server architecture has two primary advantages:

1. Distributed Application. Applications can be distributed on the network based on their requirements for resources. For example, the server might provide compute-intensive services on a system with very high computing capability, while the client runs on a workstation that provides high-end graphical display capabilities.

2. Resource sharing. A server process typically can serve many clients, so client/server architecture is a good way to implement resource sharing.

Examples of resources that are commonly shared through client/server arrangements are computing cycles, graphics capabilities, and databases. The client process often is found on an end-user node, such as a workstation or PC, while the server process often runs on more powerful systems, such as a network file server. As the LAN market has matured, many different client/server protocols have emerged, often competing with one another. For example, Apple's AppleTalk Filing Protocol (AFP) was designed to service Macintosh clients, and Sun Microsystem's Network File System (NFS) has emerged as the standard for servicing UNIX clients.

Client/server protocols are usually developed for a specific workstation environment, like Apple's AFP for the Macintosh. As the LAN marketplace has matured, however, vendors have increasingly moved toward support of multiple client/server protocols, as we will see in this chapter.

Client Requests and Remote services provided by a server typically fall into one of the categories listed below:

1. Application Access. Applications can be invoked from the client to be executed remotely on a server node. Client applications use APIs embedded in remote procedure calls (see below) to gain access to server applications.

2. Database Access. Database requests from client to server are typically made using SQL syntax, an industry standard database query language used by many vendors.

3. Print Services. Clients generate print requests which are serviced by a print server. Jobs are queued up by the print server and the client is notified when the print job has been completed.

4. Fax Services. Clients generate fax requests which are serviced by the server in a manner similar to print requests.

5. Window Services. The network operating system (NOS) typically provides software on the client workstation to pop-up windows for status messages from remote servers.

6. Network Communication. Clients access the network through APIs which use communications protocols such as IPX, TCP/IP, Ethernet, Token Ring, and others. Applications can exchange files and send messages between remote applications using these services.

Remote Procedure Call (RPC)

As increasingly advanced network applications have been developed during the past few years, an important programming technology known as Remote Procedure Call has emerged to support their development. RPC technology was used as a basis for the Network File System (NFS) by Sun Microsystems.

Understanding remote procedure calls requires that you understand what a local procedure call is, so we will explain that first (See Figure 5-2). A program consists of a sequence of instructions to the computer. As programs are developed, programmers often find that certain calculations or manipulations occur repetitively in the program. For example, a program might access a database at many different points in the program. Rather than having to recode the database access instructions over and over again, all programming languages allow a programmer to collect them into what is called a subroutine or procedure. The programmer can then write a subroutine call or procedure call statement that executes the instructions. The procedure call statement can provide parameters to the subroutine. The parameters tell the subroutine how to behave for that particular request and also provide a place for the subroutine to store the result.

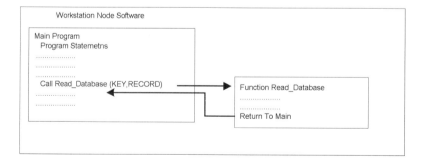

Figure 5-2. Typical Program Function Call

For example, as illustrated in the diagram, a programmer might create a subroutine called Read_Database, which is called each time it is necessary to read a record out of the database. The subroutine might use parameters such as KEY and RECORD, with KEY providing the key of the record to be read and RECORD providing space to store the record that is read from the database.

The subroutines can be part of the source code of the program or separate code which is combined with the main program after they have both been compiled. The effect is the same.

Programs invoke most of the services of the operating system and virtually all of the services of a communications subsystem through procedure calls.

Suppose that the programmer wants to make our example program into a client/server application, with the main application running on the workstation node and the database access taking place on a network server node. How can that be done?

RPC provides a method that is very attractive from the programmer's point of view. As the name implies, remote procedure calls allow subroutines such as Read_Database to be called remotely, that is, across the network, as illustrated in Figure 5-3.

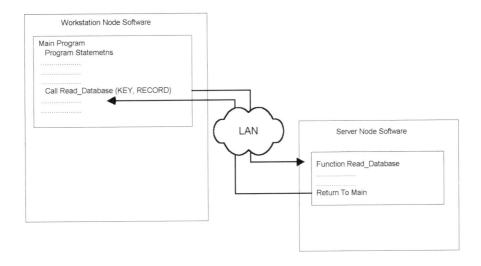

Figure 5-3. Remote Procedure Call

There are two advantages to the programmer. First of all, simplicity—the interface to the remote database is practically identical to the interface to a local database. Secondly, minimal changes to the program—RPC hides much of the complexity of the network from the programmer.

Although RPC is an elegantly simple concept, its implementation is more difficult than it appears. Procedure calls to local procedures can be assumed to be 100 percent reliable. If the procedure fails, so will the main program (if the workstation node has an equipment failure, both the main program and the procedure will stop). The same assumption cannot be made for remote procedures. The RPC implementation must take into account the delay in sending and receiving request/reply pairs across a relatively slow communications link, the possibility that the server node might fail after receiving and acknowledging the request, but before sending a reply, the possibility of getting two replies to one request, and so on.

RPC is a programming technology which has made significant inroads in data communications development. You are sure to encounter applications and network tools which employ RPC, and we will refer to its use in the following sections of this chapter. Standards for RPC protocols continue to be developed. RPC tools are available from a variety of sources, including Sun Microsystems and IBM. One widely used RPC is SunSoft's Open Network Computing (ONC) Remote Procedure Call/External Data Representation protocol.

NOS Architectures and the OSI Model

The software that manages server operations and provides services to clients is called the Network Operating System (NOS). The NOS manages the interface between the network's underlying transport capabilities and the applications resident on the server (See Figure 5-4).

As we discuss each of the NOSs, we'll compare the layered architecture of each vender solution to the OSI Model and describe the protocols supported. In general, you will find that the NOS architectures map to the OSI model as shown.

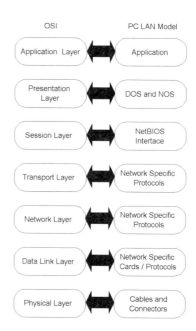

Figure 5-4. PC LAN Model and the OSI Model

Client/Server Terminology

You will come across several terms when evaluating and/or implementing Client/Server technology. Some of the terms are:

Multitasking—The ability of a computer to execute multiple processes and applications at the same time. Although a computer with a single processor can only execute one instruction at a time, a multitasking operating system can load and manage the execution of multiple applications by allocating computer processing cycles to each application in sequence. The perceived result is the simultaneous processing of multiple applications or tasks. There are two kinds of multitasking: preemptive and cooperative. With preemptive multitasking, the operating system is in charge and manages system resource allocation and task scheduling. With cooperative multitasking, applications are in charge and share resources.

Multiprocessing—In a multiprocessing environment, multiple computers are used to process a single application.

Multithreading—A thread is a process within an application that executes a specific operation. A computer capable of multithreading is one which supports multiple threads, essentially allowing applications to multitask within themselves.

Shared Memory—Allocation of a block of computer memory to be used for updating and sharing shared files for multiple client access.

Protected Memory—Isolates NOS applications and keeps one application from affecting the operation of another application.

Client/Server Environment

Listed below are the primary client operating systems and applications found in today's networking environment.

Example Client Operating Systems	Example Client Applications
DOS	E-mail
UNIX	Terminal Emulation
Windows	Database Query Applications (SQL and DB2)
OS/2	Communications Software
Apple	
Motif	

Example Server Hardware Types
Mainframe: IBM, Amdahl, Hitachi
Minicomputer: HP, SUN, DEC
Microcomputer: Compaq, IBM, HP

Example Server Operating Systems
Mainframe: MVS, AIX
Minicomputer: Unix, VMS
Microcomputer: NT, Apple, Unix, OS/2

Example Server Applications
File Sharing
Print Sharing
Communications (E-mail, Remote Access, Fax services)
Network Management (SNMP)

Network Operating System (NOS)

LAN operating systems, called Network Operating Systems (NOS), can be divided into four components (See Figure 5-5):

1. Server platform
2. Network services software
3. Network redirection software
4. Communications software

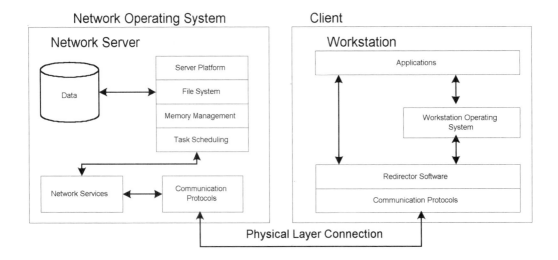

Figure 5-5. Network Operating System Platform

These components work together to support the distribution of network services to users. The relationships among the components are illustrated in Figure 5-4.

The server platform supports basic network operations, such as the network file system, memory management, and scheduling of tasks.

The network services software running on the server platform provides the user with services that range from basic (file and record locking) to very complex (database queries).

The network redirection software coexists with the operating system-DOS, OS/2, or the Macintosh operating system—in the user's workstation or PC. Applications access network services through this software.

The communications software provides the protocols needed to transmit requests for services over the network. The server receives these requests, processes them, and sends replies back to the requesting PC or workstation using the same communications software.

Server security is another important aspect of a NOS. The server should be capable of restricting access to applications and data using login IDs or some other form of control.

While these processes are typical of all NOSs, their functionality, reliability, and performance can vary significantly because of architecture differences.

XEROX NETWORK SYSTEM

We need to take a brief look at the Xerox Network Systems XNS model developed at Palo Alto Research Center (Parc), because several of the major LAN vendors (notably Novell, 3Com, and Banyan) based portions of their NOS on the XNS model (See Figure 5-6).

Xerox used a very early version of Ethernet in the 1970s to develop internal networks to support researchers experimenting with software development tools residing on personal workstations. A distributed computing system was created in which users on workstations (clients) communicated with servers of various kinds, including print and file servers.

An important part of this work was the development of network protocols to support client/server interactions over the Internet. Parc produced an Internet protocol hierarchy based on the Parc Universal Packet (PUP), a datagram-based packet format that supported encapsulating information from one layer to the next.

As you can see, the XNS protocol suite maps closely to the OSI model. At the Physical and Data Link Layers, XNS provides multilevel protocol support, including Ethernet, X.25, and leased lines.

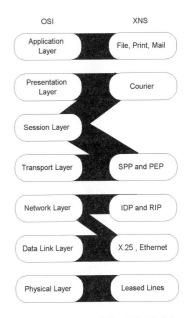

Figure 5-6. XNS and the OSI Model

At the Network Layer are the Internet Datagram Protocol (recall that datagrams are considered unreliable services, since they provide no guarantee of packet delivery or sequencing) and the Routing Information Protocol (RIP), which keeps track of network addresses and the most efficient paths for routing packets.

The Transport Layer provides the reliability that the Network Layer lacks, supporting the Sequenced Packet Protocol (SPP) for connection-oriented requirements. Applications that do not require high reliability can use the Packet Exchange Protocol (PEP), which requires less overhead.

The Session and Presentation Layer functions are provided by the Courier protocols, which are essentially Remote Procedure Call (RPC) protocols.

As we mentioned, you will see elements of the XNS architecture in several of the PC LAN architectures that will be discussed.

XNS Protocols

The XNS PC LAN architecture defines the following protocols which are used to transfer information across a network in a reliable manner. These protocols include:

- Internet Datagram Protocol (IDP)
- Routing Information Protocol (RIP)
- Sequenced Packet Protocol (SPP)
- Packet Exchange Protocol (PEP)
- Error Control Protocol
- Echo Protocol

The fundamental protocol element of the XNS PC LAN architecture is the Internet Datagram, also called an Internet Packet. Figure 5-7 depicts the format of the Internet Datagram. This network layer datagram is divided into two sections, the header and the data. The data consists of information from the layer above, the transport layer.

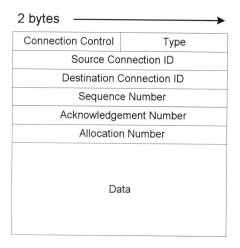

Figure 5-7. Internet Datagram Format

The header contains the protocol information used by every source, intermediate and destination node, in the network. The header consists of the following fields:

- Checksum. Used for error checking the datagram
- Length. Defines the length of the datagram in octets
- Transport Control. Contains the number of intermediate nodes passed through (hops)
- Packet Type. Defines the Transport Layer protocol used. Possible values include 1=RIP, 2=Echo, 3=Error, 4=PEP, 5=SPP
- Destination Network. Address of the network where the receiving node resides
- Destination Host. Address of the Host (node) who will receive the data
- Destination Socket. Contains virtual circuit information
- Source Network. Address of the network where the sending node resides
- Source Host. Address of the Host who is sending the data
- Source Socket. Sender virtual circuit information

The RIP protocol is also a network layer protocol. The purpose of the RIP protocol is to keep routing tables current. Routing tables contain the information needed to get datagrams through the network. Periodically, routers send out RIP datagrams to access current network routing information.

Figure 5-8 shows one of the two XNS transport layer protocols, the Sequenced Packet Protocol (SPP). SPP provides sequencing and reliable delivery information of messages. When SPP and IDP information is being sent, the SPP Header and data are encapsulated in the IDP datagram. The SPP protocol is used for connection-oriented delivery. The SPP header contains the following fields:

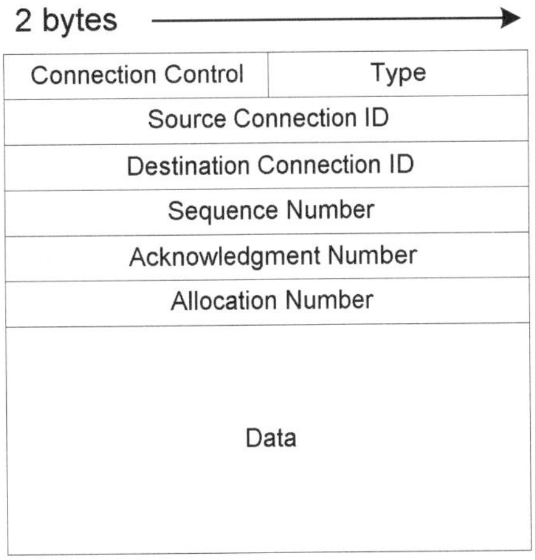

Figure 5-8. SPP Protocol

2 bytes ──────────────────────────▶	
Connection Control	Type
Source Connection ID	
Destination Connection ID	
Sequence Number	
Acknowledgment Number	
Allocation Number	
Data	

- Connection Control. Transport layer control information
- Type. Information from upper layer
- Source Connection ID. Contains the port number of the sending process
- Destination Connection ID. Contains the port number of the receiving process
- Sequence Number. Keeps up with the sequence of packets being sent
- Acknowledgment Number. The number of the next packet expected (piggybacking)
- Allocation Number. Used for flow control

The echo protocol is used as a loopback facility for sending data to a host. Data is sent to a remote host, which returns the exact data to the sending host. The error protocol is used to standardize the way communication errors are reported.

XNS Applications

XNS also defines standards for application-level functions. These include the following services:

- Clearinghouse. Provides data (such as addresses) of network resources
- Authentication and security services
- Time stamping information
- Character code standards
- Document management and interchange
- Print, file and mail standards

NOVELL NETWARE

Novell, Inc., incorporated in 1983, has experienced phenomenal success. Novell's mainstay product is NetWare 386, but its newer products, NetWare 486 and NetWare Lite, represent the company's strategic direction for the future. Portable NetWare is designed for implementation on operating system platforms such as UNIX, OS2, and Digital's VMS. Novell's 1991 agreement with IBM, in which IBM declared NetWare a strategic networking product for the IBM environment, solidified Novell's position as an industry leader.

Originally developed as a file server operating system, NetWare has gone through several revisions, each one strengthening and broadening the product. Today, it provides a full suite of networking functions and services, including communications, database, and message store-and-forward. With NetWare, customers can start small and build to massive networks, based on their networking requirements.

Novell's distributed, multitasking Network Operating System (NOS) is designed to provide and coordinate all network services, including file directory services, print services, software protection services, network security, and messaging. PCs and workstations on the network use the Advanced NetWare Core Protocol (NCP) to communicate with the operating system to obtain network services for their local applications.

NetWare's architecture readily maps to the OSI model as shown (See Figure 5-9). At the Physical and Data Link Layers, NetWare supports Ethernet, IEEE 802.2, and IEEE 802.5, as well as Token Ring, ARCNET, and a number of other network architectures. Network and Transport Layer functions are handled by the Internetwork Packet Exchange (IPX) and Sequenced Packet Exchange (SPX), both of which are variations of XNS protocols.

At the Session Layer, the Network Basic Input Output System (NetBIOS) interface, developed by Sytek for IBM, is supported. The Presentation and Application Layers provide NetWare Core Protocol services, PC-DOS services, additional NetWare value-added services, and various user applications.

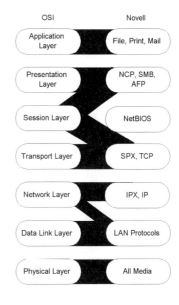

Figure 5-9. Novell NetWare and the OSI Model

NetWare Client

NetWare supports many types of clients, such as a DOS, Windows, Apple, and OS/2 computers. There are essentially six software components needed on a client workstation:

1. Application Software
2. Local Operating System
3. Redirector or Requester
4. Communications Protocol Stack
5. Link Layer Support Software
6. LAN Driver

The NetWare client software runs in conjunction with the client operating system to provide the client access to NetWare servers and server resources. The NetWare Shell (also referred to as the redirector) is the software that surrounds the local operating system and intercepts requests to the operating system. It determines if the request can be serviced locally or needs to be serviced across a network. File services and print services are primary examples of requests handled by the NetWare shell. NETX.COM is the TSR (Terminate and Stay Resident in memory) software that "redirects" application requests to the local or remote operating system.

Newer NetWare implementations no longer use the NetWare shell. Instead, the DOS Requester is used. The DOS requester contains VLMs (Virtual Loadable Modules) which are software components used to carry out the same general functions as the NetWare Shell. Examples of VLMs are:

- IPXNCP.VLM: handles NCP and IPX services
- NDS.VLM: used for 4.x NetWare Directory Services
- FIO.VLM: used for file input/output when requesting network resources

A sample network request from a client would be processed as follows (Refer to Figure 5-10):

1. The NetWare Shell determines whether the request can be serviced locally or across the network.
2. If the request can be handled locally it is serviced by the local operating system (or) if the request is for network resources, it is converted to the appropriate format (such as NCP).
3. The request is put into an IPX packet to be transmitted.
4. The packet is passed to the NIC in the appropriate format (Ethernet, Token Ring, etc.).
5. The packet is routed to the appropriate server process.

NetWare Server

The NetWare operating system resides on the server hardware platform. The hardware platform typically consists of a 486 or better, along with server peripherals such as printers, disks, and modems. The NetWare server also contains NICs that attach the server to the network. The NetWare NOS is a multitasking operating system which includes services such as starting a program, memory management, task scheduling, and interprocess communication, as well as communication across a network. The various programs that run on the server are referred to as Network Loadable Modules (NLMs).

Clients attached to the network request various services from the NetWare server. The NCP protocol is normally used to request such services. Examples of NetWare server and underlying NCP services are listed below:

1. Bindery services: The bindery database file contains definitions for users, groups, and workgroups used by NetWare for versions 3.x and below. With NetWare 4.x, the NetWare directory database is used instead (NDS). These services include adding, deleting, and modifying user data such as passwords and file access.

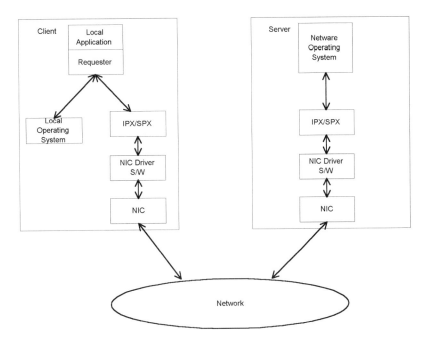

Figure 5-10. Example Client Request

2. Connection services: All computers, applications, processes and resources attached to a NetWare network require a connection number. Connection services provide management of these connections.

3. File Services: Contains functions to create, open, close, and delete files.

4. Message Services: This service provides the capability to send messages to various users or groups of users over the network. Broadcast messages over the LAN are sent using the NCP message services.

5. Print Services: This service provides management of print processes such as print spooling and queuing.

NetWare Products

Novell has a broad range of products to meet customers needs. Examples of Novell products include the following:

Entry-Level Solution (ELS) NetWare Level I

Designed for small businesses and workgroups. ELS NetWare 1 lets up to four users share files and resources. The server is nondedicated. This entry-level product does not support O/S2, Macintoshes, or internetworking. This level is no longer supported by Novell.

Entry-Level Solution (ELS) NetWare Level II

Still targeted at small businesses and workgroups. ELS NetWare II lets up to eight users share files and resources. The server can be either dedicated or nondedicated. This entry-level product supports OS2, Macintoshes, and has limited internetworking capabilities. This level is no longer supported by Novell.

Advanced NetWare 286 V2. 15

For medium to large companies and workgroups in large companies. The server can be either dedicated or nondedicated. Advanced NetWare 286 supports up to 100 concurrent users and provides extensive internetworking capabilities. This level is no longer supported by Novell.

System Fault Tolerant (SFT) NetWare 286 V2.15

Novell's high-end, 286-based product. It provides all of the functionality of Advanced NetWare 286 V2.15 plus data redundancy, duplexed channels, mirrored disk drives, and transaction tracking. SFT NetWare requires a dedicated server. This level is no longer supported by Novell.

Host Connectivity

In the host gateway category, Novell provides SNA gateways to IBM hosts (3270, 5250, and LU6.2) and asynchronous and X.25 gateways for connections to asynchronous hosts. Wide area networking products connect remote networks over a variety of WAN options, from asynchronous dial-up lines to the X.25 public data network to high-performance T-1 options.

TCP/IP Support

In 1989, Novell merged with Excelan, a leading TCP/IP networking vendor. This partnership has allowed Novell to expand its support of TCP/IP networks, offering TCP connectivity products for DOS, OS2, Macintosh, and UNIX (LAN WorkPlace products).

NetWare 386 versions 3.11 and 3.12

These releases continue to be Novell's most popular products. NetWare 386 provides all of the functionality available in the 286 product plus extensive internetworking. In NetWare 386, Novell entirely redesigned the Network Operating System to improve performance, reliability, and security, to support an open architecture, and to address important technology trends and market demands. These include:

- Support of multiple media and transport protocols

- Support of multivendor, enterprise-wide computing environments, such as IBM's MVS and VM, Digital's VMS, and generic UNIX for host systems through Portable NetWare

- Support of multiple desktop computing environments, including DOS, O/S2, Macintosh, and UNIX workstations

- Support of multiple client/server LAN models, including IBM and Microsoft's Server Message Block (SMB), implemented in LAN Manager, Apple's AppleTalk Filing Protocol (AFP), and Sun Microsystem's Network File System (NFS)

Novell has also responded to several weaknesses in NetWare consistently pointed out by customers, including the absence of a universal naming convention, inflexible print services, and the difficulty of installing and setting up the network. All of these criticisms were dealt with in NetWare 386.

NetWare 4.x. This version includes enhancements to version 3.x. These enhanced features include:

1. NetWare Directory Services (NDS). NDS improves internetworking management by providing a graphical user interface that network administrators can use to manage network resources.

2. Enhanced security features.

3. Enhanced language support.

4. High Capacity storage feature. HCS allows for the integration of optical disks or tapes into the NetWare Filing System.

NetWare Lite

NetWare Lite is Novell's low-end LAN solution for small businesses. It supports workgroup applications for twenty-five or fewer workstations. It has file and printer sharing capabilities as well as limited security features. NetWare Lite is a peer-to-peer network only; dedicated servers are not supported.

Personal NetWare

Personal NetWare supports up to fifty nodes per server and up to fifty interconnected servers. It is fully compatible with other versions of NetWare and supports SNMP standards.

NetWare SFT III

In September 1994, Novell announced NetWare System Fault Tolerance (SFT) technology, Novell's fault-tolerance solution for networks, with NetWare release 4.1. The mirrored-server technology of NetWare SFT III is specifically designed for those businesses that require critical applications and data to be continuously available. SFT III avoids the problems related to network downtime by maintaining two servers with identical memory images and disk contents. If one server fails, the other server automatically takes over. This technology allows businesses to maintain data and application reliablilty with far less expensive and complex hardware than was previously possible.

NetWare NFS

NetWare NFS (Network File System) links UNIX systems to the NetWare environment. The NetWare NFS NetWare Loadable Modules (NLMs) provide NFS and FTP (File Transfer Protocol) services to UNIX clients. NetWare NFS also provides a two-directional print gateway between NetWare and UNIX and, like NetWare/IP, includes the X-Window application XCONSOLE to allow X-Window System users to administer the NetWare server remotely.

NetWare NFS Gateway

NFS Gateway is a product which allows DOS and Windows clients on a NetWare network to access NFS files on a UNIX system as easily as they access those on NetWare servers.

LAN WorkPlace for DOS

This product gives DOS and MS Windows users easy access to UNIX systems and other TCP/IP network resources, with or without NetWare. For large NetWare networks requiring this functionality, Novell provides its LAN WorkGroup product. In mid-1994, Novell announced a free Internet Access Toolkit with the purchase of LAN WorkPlace.

LAN WorkGroup

LAN WorkGroup, Novell's latest WorkPlace product, is a server product providing NetWare users with transparent access to a number of TCP/IP network resources through both DOS and Windows interfaces. This gives NetWare users easy access to UNIX systems and other hosts from the desktop. LAN WorkGroup also has network management features, such as the automatic configuration and assignment of IP addresses for each workstation.

NetWare FLeX/IP

This software is a set of TCP/IP utilities providing connectivity between UNIX systems and the NetWare environment. The package provides a two-way print gateway between UNIX and NetWare environments allowing UNIX users to transfer files to and from the NetWare environment. It also allows remote management of NetWare servers from an X Window System using the XCONSOLE application.

Network Management

Novell's network and systems management products focus on three areas of management: monitoring and controlling the infrastructure, managing system services and administration facilities, and allowing network management to the desktop.

NetWare Protocols

The networking protocols used to handle requests over the network in the example above are NCP, SPX and/or IPX, ODI, and the underlying data link layer protocols such as Ethernet and Token Ring.

Figure 5-11 depicts the IPX and SPX protocol details. Note that the IPX protocol format is identical to the XNS Internetwork Datagram Protocol (IDP). The protocol fields are described below. This figure also shows how this data is contained within an Ethernet MAC frame.

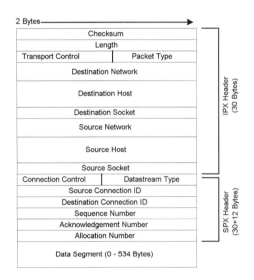

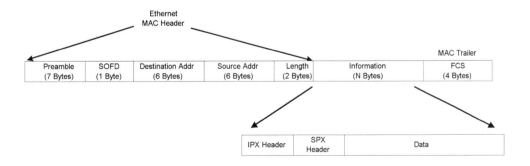

Figure 5-11. SPX/IPX Protocols

Internet Packet Exchange (IPX)

- Checksum: set to FFFF (hex) and is not used
- Length: gives length of datagram (0-576 bytes)
- Transport Control Field: used to track the "hop count"
- Packet Type: (defined by Xerox) contains values relative to RIP, PEP, SPP, NCP, and other options
- Destination Network: subnetwork address
- Destination Host: physical hardware address of destination node
- Destination Socket: (defined by Xerox) defines the process within the destination node such as error handling, routing information, and so on
- Source Network: assigned by network administrator
- Source Host: address of process within destination host

Sequenced Packet Exchange (SPX)

- Connection Control: contains flags which control the data flow source and destination
- Datastream Type: identifies the type of data within the packet
- Source Connection ID: identifies the connection within the source host (process)
- Destination Connection ID: identifies the connection within the destination host (process)
- Sequence Number: provides the packet number of the packet being sent
- Acknowledge Number: provides the packet number of the packet expected from the other node
- Data Field: contains up to 534 bytes of data from the upper layers

The Open Data-Link Interface (ODI) Specification

The ODI specification (See Figure 5-12) was developed by Novell and Apple in 1989. The purpose of ODI is to seamlessly integrate data link layer protocols with upper layer protocols. This allows multiple network layer protocols to access the network through one or more NICs.

ODI consists of three major components:

Multiple Link Interface Drivers
Each MLID device driver is unique to the underlying adapter hardware and the LAN media (Ethernet, Token Ring, etc.). MLIDs control the communication between the LSL and the LAN media. MLID strips the header and trailer from the MAC frame and passes the data to the LSL layer.

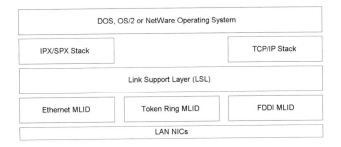

Figure 5-12. Open Data-Link Interface

Link Support Layer (LSL)
LSL handles the communication between the MLID and the upper layer protocol stacks. It is the heart of the ODI specification.

Protocol Stacks
Upper layer protocol stacks include IPX/SPX, AppleTalk, and TCP/IP.

For a NetWare DOS client to access server resources using ODI, the following files are needed:

LSL.COM: LSL Driver
MLID.COM: Specific MLID driver
IPXODI.COM: IPX Protocol Stack
NETX.COM: NetWare Shell
    (or)
*.VLM: NetWare Virtual Loadable Modules

Netware Core Protocol

The NCP protocol is used to send data to and from the client and server. The NCP protocol provides request/reply pairs for functions such as printing, file access, and security. For example, when reading and writing files to a network server, NCP would provide the following:

- Creating or breaking a connection
- User validation
- Mapping of appropriate disk drives
- File rights validation
- File access

BANYAN VINES

Banyan Systems, Inc., founded in 1983, is one of the unsung heroes among LAN vendors, offering a broad range of products that provide sophisticated services in wide area network environments. Banyan built its Network Operating System, VINES (Virtual Networking System), as a layer on top of UNIX in order to take advantage of the operating system's multiuser, multitasking capabilities (See Figure 5-13). But the company made this technological leap before most large companies believed in the viability of UNIX as an enterprise-level operating system.

PC clustering, wide area networking, and distributed network applications are the major strengths of Banyan's networking strategy. VINES runs on multiple hardware platforms, including Banyan's own line of network servers and PC compatibles. With VINES, dissimilar networks can coexist in the same network environment. Macintoshes can also be integrated into the network; they can be directly connected to a server via Ethernet or using an Ethernet-to-LocalTalk bridge.

VINES supports file sharing, disk sharing, printer sharing, electronic mail messaging, synchronous and asynchronous communications, file backup and recovery, and internetworking.

An aspect of VINES that sets it apart from other LAN vendors is StreetTalk, the VINES global naming and file directory service. StreetTalk simplifies the task of finding paths through networks by assigning three-part names to every network node (for example, Printer@Sales@London). Resources (like a printer) can easily be accessed by the user, because the location of a resource does not matter if the user knows the resources StreetTalk "name." Each group (sales in this example) occupies a single server. Organizations (London) are logical groupings that are not permanently tied to physical resources.

The lower protocols of the Banyan solution support all of the major LAN and WAN connections. The Physical Layer supports baseband and broadband LAN backbones, as well as asynchronous and synchronous communications interfaces for remote access. The Data Link Layer includes IEEE 802.3 and 802.5, ARCNET, and IBM's SDLC and BSC protocols. The VINES Fragmentation Protocol (VFRP) lies between the Data Link and Network Layers to segment packets and to reassemble them into frames as necessary.

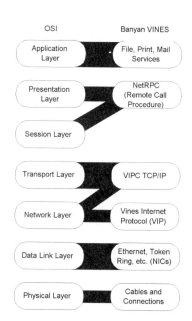

Figure 5-13. Banyan Vines and the OSI Model

The Network Layer provides datagram services using both industry standard protocols and Banyan's own proprietary protocols. VINES provides the Internet Protocol (IP), Internet Control Message Protocol (ICMP), and Address Resolution Protocol (ARP) at the Network Layer.

Transmission Control Protocol (TCP) and User Datagram Protocol (UDP) are supported at the Transport Layer. Banyan's proprietary protocols include the VINES Interprocess Communications Protocol (VIPC) for both datagram (unreliable) and reliable message service, and the VINES Sequenced Packet Protocol (VSPP) for data stream service (both of these protocols are based on XNS protocols). The Session and Transport Layers are implemented using the VINES NetRPC utility, a derivative of XNS Courier.

The VINES Application Layer provides file and print services, electronic mail, and support for UNIX and DOS applications. It also supports a NetBIOS emulation.

Host Communications Features

Banyan's 3270/SNA and 3270/BSC emulation products let PCs and VINES servers emulate IBM 3270-type devices. Banyan also provides asynchronous communication capabilities through its Asynchronous Terminal Emulation program, which lets PCs emulate such terminals as DEC's VT/100 and VT/52. The program also supports access to remote hosts.

Banyan also supports TCP/IP for connections between Banyan servers and TCP/IP networks over Ethernet and Token Ring.

The latest enhancements fall into two categories: those which relate to the enterprise network services, which are an integral part of VINES, and those which relate to the VINES NOS itself.

Multiple Platform Support

VINES 6.0 includes integration with NetWare, so that a VINES 6.0 server can act as the host of the ENS for NetWare option, thereby allowing a single server to provide services for both VINES and NetWare users. A single VINES/ENS server can host up to eight NetWare servers, and as part of the ENS for NetWare option, a VINES server can connect a single enterprise client to file and print services on VINES, NetWare, and ENS for UNIX servers to provide cross-platform resource sharing. VINES 6.0 also provides TCP/IP, DOS, Windows, and OS/2 workstation support.

UNIX

The UNIX operating system has been around since the early 1970s. It is responsible for managing hardware resources that exist in a computer such as memory, hard disk drives, video output, and CPU processing. It provides programmer capabilities such as editors, compilers, assemblers, and text formatters. The main components of the UNIX operating system are:

- User Processes
- File System
- Kernel
- Shell

In addition to these components, UNIX also provides networking capabilities in the base operating system. As with all client/server processes, a UNIX server waits for commands from clients and then executes the commands. UNIX is not usually included when discussing the NOS market because the networking functionality has been built into the operating system for quite some time.

Networking in UNIX is accomplished using interprocess communications (IPC) between applications and the networking protocol stacks. Most UNIX systems provide multiple protocol suites including TCP/IP, SNA, XNS, and NetBIOS. TCP/IP is normally used as the protocol stack in UNIX networking implementation. Figure 5-14 shows a typical UNIX protocol stack.

Programs are written under UNIX using Application Programming Interfaces (APIs). The three most common UNIX APIs are sockets, TLIs, and Streams.

Sockets are general purpose IPC mechanisms used to communicate between a process and a protocol stack or another process. The UNIX socket interface supports both TCP and XNS protocol implementations. The socket interface provides a set of system calls (functions) which networking applications can use to perform networking I/O.

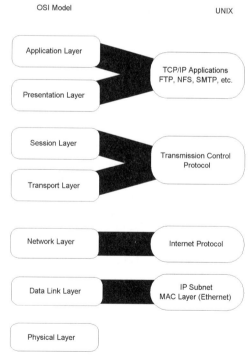

Figure 5-14. UNIX and the OSI Model

As the name implies, TLI provides an interface to the transport layer such as the Transmission Control Protocol (TCP). Applications can use standard TLI function calls to access communication protocol stacks under UNIX.

Streams are another I/O mechanism used under UNIX. Streams are low-level mechanisms used by networking protocol programmers to gain access to lower layers of a protocol stack and for creating protocol software modules.

LANTASTIC

LANtastic, developed by Artisoft, has gained popularity primarily as a small business, peer-to-peer networking solution. Artisoft's LANtastic network operating systems latest version includes Universal Client technology allowing networks to be configured as peer-to-peer, dedicated server, or a combination of both.

The LANtastic system allows the user to choose whether to dedicate a server or to run the server on a user's workstation. The LANtastic NOS can be set up so that every workstation is both a server and client; however, on networks with greater than ten nodes, a more common configuration would contain only a few servers. Typically, these in turn would be divided into file and print servers. In order to provide solutions for communication across multiple hardware and software platforms, Artisoft offers a series of connectivity solutions for support of TCP/IP, Macintosh, and OS/2. Other LANtastic strengths include its administration facilities—whereby users have freedom to manage their own workstation's resources and to define access to them. However, there is also the possibility of putting all the administration and management functions in the hands of one person. Figure 5-15 shows how the LANtastic NOS stacks up to the OSI Model.

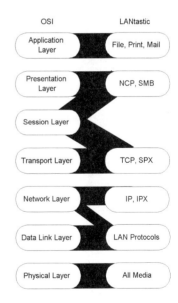

Figure 5-15. LANtastic and the OSI

LANtastic Products

Simply LANtastic

Simply LANtastic is a streamlined, entry-level version of the LANtastic NOS designed for small businesses and home offices. It is easy to set up and allows printers, hard drives, and CD-ROM drives to be shared. Users can exchange files between computers and pool network software applications and communicate via e-mail. Security is provided down to subdirectory level for full-access, read-only access, or no access. Simply LANtastic Starter Kit provides two parallel port adapters, software for two PCs, one 25" connector cable, two 12" parallel port extender cables, and two AC adapters.

LANtastic 6.0

This release brings LANtastic into line with the growing trend away from purely peer-to-peer NOSs towards the new, more scaleable Universal Client NOS for small- to medium-sized businesses. The universal client technology provides seamless connectivity to network servers of Novell Inc., Microsoft Corp., and IBM Corp. Additions for Version 6.0 include a groupware system, providing e-mail, network scheduling, faxing, and paging features. It also provides improved network performance and network management features. LANtastic does not require a dedicated server and allows any PC on the network to act as a server, workstation, or both. It supports up to 500 users per server in both DOS and Windows environments and provides a choice of a DOS character-based or graphical Windows menu systems. DOS and Windows workstations can be freely mixed, either as clients or servers. For the hardware side of the network, Artisoft manufactures the NodeRunner 2000 series of Ethernet adapter cards, and these are available for ISA, EISA, and Micro Channel PC bus architectures.

Universal Client Technology

This feature includes NetWare Core Protocol (NCP) support, which allows LANtastic 6.0 workstations to access NetWare servers for file and print services. It also means that users can operate both types of networks transparently in the same LAN while adding flexible networking features to departmental workgroups in an existing NetWare installation. Artisoft and Novell have also entered into an agreement which includes Novell certification of the interoperability between LANtastic network clients and NetWare servers. Through Server Message Block (SMB) client support, LANtastic 6.0 workstations gain the capability to access any SMB-based server for file and print services. This support means that LANtastic 6.0 network users are provided with access to servers running Microsoft Windows NT and Windows for Workgroups, IBM LAN Server, and any other system which is SMB Version 1.0 compatible.

IBM PC LANs

IBM supports two Network Operating Systems, the PC LAN Program and the OS/2 LAN Server (See Figure 5-16). The PC LAN Program is an enhancement of the earlier PC LAN Support Program, which had been criticized by customers for its poor performance. Recent improvements have resulted in better performance and expanded services.

While IBM supports the PC LAN Program, the company's clear desire, until recently, has been that customers adopt OS/2 Extended Edition. The strengths of the OS/2 LAN Server reflect that emphasis. However, the announcement in 1991 that IBM would market Novell's NetWare demonstrated continuing market pressure for IBM to support an open environment.

The PC LAN program provides a common support package for all IBM LAN technologies running under PC-DOS, including Token Ring, PC Network (broadband and baseband), and Ethernet. PC LAN Program makes communication among users available through file copy, messaging services, and resource sharing.

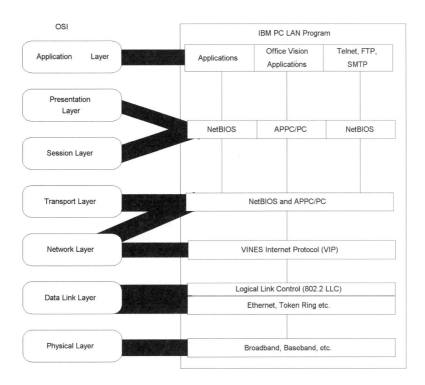

Figure 5-16. IBM PC LAN and the OSI Model

PC LAN protocols are concentrated at the Application and Presentation Layers of the OSI Model, including DOS, Redirector, and SMB elements (See Figure 5-17). The NetBIOS driver is the Session Layer interface. The PC LAN Support Program provides support at the Transport and Network Layers, and IEEE 802.3 and IEEE 802.5 are supported at the Data Link Layer.

OS/2 LAN Server extends the functionality of the PC LAN Program to provide complete networking support for OS/2 users. It is in compliance with IBM's guidelines for System Application Architecture (SAA) and supports all IBM LANs and LAN interfaces, including IEEE 802.3, NetBIOS, and LU6.2 (APPC).

At the lower layers, OS2 LAN Server depends on the PC LAN Program implementation for functionality. NetBIOS support continues at all layers, but support for LU 6.2 (APPC) is added at the Network, Transport, and Session Layers, giving OS/2 LAN Server users access to the SNA world. The Presentation and Application Layers support OS/2 Extended Edition applications.

The key difference between the PC LAN Program and OS/2 LAN Server lies in the latter's support of distributed processing applications. While the PC LAN Program was designed to support stand-alone LANs (although bridges and gateways are supported), OS/2 LAN Server is integrated into IBM's larger networking strategy. IBM is encouraging customers to adopt the company's OS/2 solution as their enterprise-wide PC networking strategy.

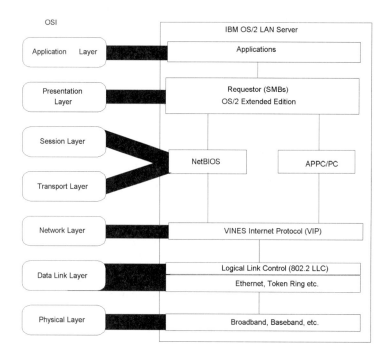

Figure 5-17. OS/2 LAN Server and the OSI Model

Host Communications Software

IBM provides a number of communications packages for exchange of information with IBM hosts. SNA Host Gateway provides remote connection between the PC network and multiple host applications; the PCs must have PC3270 Emulation Program and an SDLC adapter card installed. And SNA APPC PC lets an application running on a network PC communicate on a peer-to-peer basis with another program on another PC, System/36, or SNA/3270 host over an SDLC link.

APPLETALK VS. THE OSI MODEL

AppleTalk, Apple Computer, Inc.'s LAN for connecting Macintosh computers, has taken giant strides in the past few years. Originally, connecting Macs so they could share printers was easy. Users plugged an inexpensive LocalTalk connection box into the Mac's printer port and attached a two-meter cable to the connection box on the next Mac in the network. Each Mac was shipped with the hardware and software required to participate in the network.

However easy the network was to build, users argued that it had major flaws. Its data transfer rate was only 230.4 Kbps, and the network could only support 32 Macs. Customers were also required to use shielded twisted pair wiring.

Apple's determination to be taken seriously in the world of business resulted in AppleTalk Phase II, announced in June 1989. AppleTalk Phase II represented a complete turnaround from Apple's usual proprietary stance. It connects LocalTalk, Ethernet, Token Ring, and FDDI networks, supports coaxial cable as well as shielded twisted pair, and supports products that facilitate the integration of Macintoshes into Digital, IBM, TCP/IP, UNIX, and OSI environments. A potential 16 million nodes can be supported by the network.

Figure 5-18 shows the AppleTalk protocol stack and how it relates to the OSI Model.

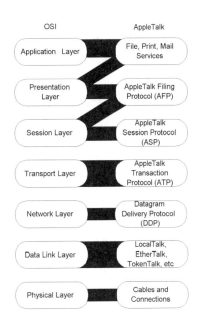

Figure 5-18. AppleTalk and the OSI Model

At the Data Link Layer, AppleTalk uses the AppleTalk Link Access Protocol (ALAP). Medium access control for bus or star topologies and twisted pair wiring is CSMA/CA (Carrier Sense Multiple Access with Collision Avoidance). CSMA/CD is used for Ethernet, and token passing is used for Token Ring environments. At the Network Layer, the Datagram Delivery Protocol (DDP) supports communication between two sockets, the addressable entities within a node. Address translation is the responsibility of the AppleTalk Address Resolution Protocol (AARP).

Several different protocols are supported in the Transport Layer. The Routing Table Maintenance Protocol (RTMP) maintains information about the current configuration of the network. The AppleTalk Echo Protocol (AEP) is used for maintenance. The Name Binding Protocol (NBP) provides translations between character-oriented names and Internet socket addresses. Reliable socket-to-socket transmissions are the responsibility of the AppleTalk Transaction Protocol (ATP).

Four protocols are also supported at the Session Layer. The AppleTalk Session Protocol (ASP) opens, maintains, and closes sessions between sockets. The AppleTalk Data Stream Protocol (ADSP) ensures reliable service between sockets. The Zone Information Protocol (ZIP) maintains a "map" of the zones within the network. The fourth protocol, the Printer Access Protocol (PAP), handles requests for access to Apple LaserWriter printers.

At the Presentation and Application Layers, the AppleTalk Filing Protocol (AFP) provides access to remote files, and the PostScript protocol supports desktop publishing.

One of the major benefits of AppleTalk networks, dynamic node addressing, continues to be supported in AppleTalk Phase II. This feature assigns node addresses dynamically when the machine is powered on.

Host Communications Software

MacDFT (which supports both DFT and CUT mode) lets a Macintosh emulate an IBM 327x terminal to access data on an IBM host running VM/CMS or MVS/TSO. MacAPPC lets Macintosh computers exchange information with other systems supporting IBM's Advanced Program-to-Program Communications (APPC) on a peer-to-peer basis.

The AppleTalk Internet Router lets Macintoshes communicate with systems on an Internet, and MacX25 is available for communications over an X.25 packet switched network. Apple also provides AppleTalk for VMS and MacTCP, programs which developers can use to create Macintosh applications for communication with DEC systems and TCP/IP environments.

WINDOWS NT

Windows NT is Microsoft Corporation's 32-bit operating system. Microsoft has positioned Windows NT as a mission-critical, platform independent operating system. It provides native application support for MS-DOS, 16-bit and 32-bit Windows, Windows 95, and POSIX 1003.1 based applications. Figure 5-19 shows how Windows NT stacks up to the OSI Model.

At the Physical and Data Link Layers of the OSI model are the network adapter card drivers that connect Windows NT to the related network adapter card. These layers also include the Remote Access Service drivers that allow remote network access.

Above these drivers is the NDIS 3.0 Interface, described later in this chapter.

The transport protocol drivers reside in the Network and Transport Layers of the OSI model. On top of these transport protocol drivers, in the Transport Layer of the OSI model, is the Transport Driver Interface.

The transport protocols pass data to the network adapter card drivers through the NDIS 3.0 Interface, and communicate with the redirector via the Transport Driver Interface.

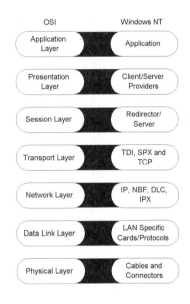

Figure 5-19. Windows NT and the OSI Model

Two software components, redirectors and servers, reside at the Session Layer of the OSI model. These components are implemented as file system drivers, which enable client applications to call a single API to request file access without the need to know whether the file is local or remote. Redirectors handle the client side functions, whereas servers handle the server side processing. Redirectors and servers communicate with the protocol stacks to which they are bound via the TDI. Additional redirectors are necessary to communicate with non-Microsoft networks.

An example of how the redirector works is shown in Figure 5-20. When a user process wants to open a file, it passes the request to the I/O Manager executive service. The I/O Manager then recognizes that the file physically resides on another computer and passes the request to the redirector. The redirector then uses the TDI to pass the request down to the network adapter card and on to the appropriate server computer for processing.

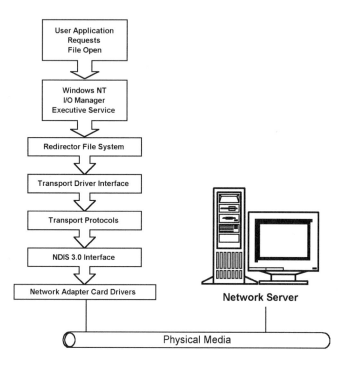

Figure 5-20. NT Redirector

Server components handle the requests from client-side redirectors and allow access to the local resources of the server computer. Figure 5-21 illustrates a typical file read request passed from a client redirector. When a client redirector passes the file read request to the remote computer, the low-level network drivers receive the request and pass it to the server driver. The server driver passes the request on to the appropriate local file system driver, which retrieves the data from the disk drive and passes it back to the server driver. The server driver then forwards the data to the low-level network drivers for transmission to the client computer.

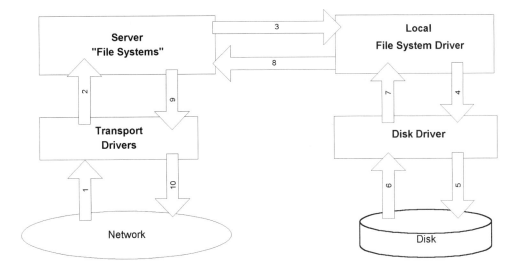

Figure 21. Windows NT Server

At the Presentation Layer of the OSI Model reside the various executive services of the operating system. The executive services include the IO Manager, Object Manager, Virtual Memory Manager, and other services.

The Application Layer of the OSI model encompasses software called providers. A provider is a component that enables Windows NT to communicate with the network. Windows NT includes providers for Windows NT networks, Client Services for NetWare, and Gateway Services for NetWare (NT Server). The various network vendors supply the provider software necessary to connect Windows NT to their networks.

On top of the various provider components sits a Multiple Provider Router (MPR). The MPR provides a single API for applications and ensures that file requests are sent to the appropriate file systems or redirectors as necessary.

Windows NT is available in two versions, Workstation and Server. The workstation version is targeted as a high-end desktop PC operating system, whereas the server version is targeted as a network file, print, and application server (See the Figure 5-22).

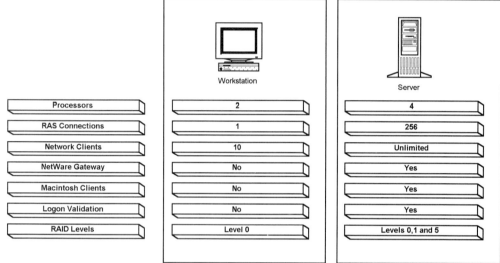

	Workstation	Server
Processors	2	4
RAS Connections	1	256
Network Clients	10	Unlimited
NetWare Gateway	No	Yes
Macintosh Clients	No	Yes
Logon Validation	No	Yes
RAID Levels	Level 0	Levels 0,1 and 5

Figure 5-22. Windows NT Feature Comparison

Windows NT features include:

- Multitasking operating system
- Multiple hardware platform support
- Network support
- Security
- Multiple file system support
- Disk Management utilities

Windows NT Server adds the following functionality:

- Greater NetWare support
- Centralized user management
- Disk fault tolerance

Multitasking Operating System
Windows NT is a full 32-bit preemptive multitasking operating system. Processes are executed in separate memory spaces, preventing any one process from crashing the operating system. Each process can run multiple threads of execution. Each thread of execution is scheduled for processing on the processors available to the operating system.

Symmetric Multi-Processing (SMP) allows Windows NT to operate on systems with more than one central processor. Out of the box, Windows NT Workstation supports two processors, and Windows NT Server supports four processors. Some OEM implementations of NT Server support up to thirty-two processors. When multiple processors are available to Windows NT, the operating system is capable of executing process threads simultaneously, which can significantly increase system throughput.

Multiple Hardware Platform Support
Versions of Windows NT are available for the following hardware platforms:

- Intel 386DX/486, Pentium, Pentium Pro
- MIPS R4000
- DEC Alpha AXP
- Intergraph Clipper
- PowerPC

Windows NT was designed with platform portability as a central feature, utilizing a hardware abstraction layer to isolate physical hardware from the operating system. Therefore, portability to new hardware platforms of the future is virtually assured.

Network Support
Networking was designed into NT from the ground up. Unlike MS-DOS based network operating systems, which are built on top of DOS, Windows NT's networking is incorporated at the heart of the operating system, utilizing 32-bit network and protocol drivers. NT supports fifteen network protocols, including Client Services for NetWare, TCP/IP, NetBEUI, IPX/SPX, DCE, RPC, and DLC.

Windows NT supports the following networks:

- Banyan VINES
- DEC PATHWORKS
- IBM LAN Server
- Microsoft LAN Manager
- Novell NetWare
- Internet

Windows NT Server adds the support for Macintosh clients and Gateway services for NetWare. The gateway services for NetWare allow clients on the NT Network to access file and print resources on NetWare networks, without the NT clients need for running the IPX/SPX protocol. File and Print Services for NetWare allows NetWare clients to access file and print resources of the NT Server, making the NT Server visible to NetWare clients.

Windows NT Workstation allows ten inbound client network connections, whereas NT Server allows unlimited inbound client network connections.

Windows NT Server uses a domain administration model that allows multiple servers to be administered as if they were a single server. User logins only need to be configured on one server, rather than at each server in the domain.

For remote network connectivity, Remote Access Services (RAS) enable both inbound and outbound connections via ISDN, X.25, and standard phone lines. With RAS, NT computers can dial-out to connect to remote servers, and also allow dial-in connections from other workstations. NT Workstation allows one connection. NT Server allows up to 256 simultaneous RAS connections.

Security

Windows NT includes sophisticated security features to control access to system resources. These security features include the following:

- Certified C2 level security
- File and directory level security with NTFS file system
- Local desktop security (NT Workstation); user login required for access
- Centralized management of network logins (NT Server)
- Account lockout capabilities to prevent unlimited login attempts

Windows NT uses a resource/permission model for security, whereby users (or groups of users) are given permissions to access various system resources.

Multiple File System Support

Windows NT supports the following file systems:

- FAT   (File Allocation Table—MS-DOS)
- HPFS   (High Performance File System—OS/2)
- NTFS   (NT Files System—Windows NT)

While the FAT and HPFS file systems are supported for compatibility with MS-DOS and OS/2, NTFS is the file system of choice, due to its recoverability and security. NTFS is a transaction based file system which logs all directory and file updates. In case of a system failure, power loss, etc., the NTFS logged information permits undo/redo operations, allowing the file system to recover lost data. Long file names are allowed on all of the supported file systems.

Disk Management and Fault Tolerance

Windows NT Workstation provides enhanced disk management ability. Volume Sets allow non-contiguous areas of free space on a single hard disk to be combined into a single logical drive. Stripe Sets allow areas of free space on multiple hard disks to be combined into a single logical drive. Both volume sets and stripe sets are used to accomplish the same goal of combining unused disk space into single logical drive units.

When used with the NTFS file system, a volume set can be extended to include additional disk space without the need to reformat the existing volume.

Windows NT Server adds the following fault tolerance capabilities to the disk management features found in NT Workstation:

- Disk mirroring (RAID Level 1)
- Disk striping with parity (RAID Level 5)
- Sector sparing

Disk mirroring maintains a backup copy of a partition. Mirroring requires a minimum of two hard disks. It is the least expensive way to add fault tolerance and data redundancy to an NT system. However, because it requires duplicate disks, it is not the most cost-effective method. Additionally, disk duplexing may be used, which utilizes additional disk controllers. This has the benefit of increased performance and guards against controller failures as well as media failures.

Striping with parity writes disk data across an array of disks and maintains parity information. If a disk in the array fails, the system can recover the lost data from the remaining disk data and parity information. While read performance is typically very good, write performance is slower due to the need for writing to an array of disks. Also, in case of a failure, read performance slows, since the system must recover the lost data using the remaining valid data. Because this is a software based solution, it is not as fast as a hardware based disk array.

Sector sparing is used in addition to the RAID technology, and is generally available with SCSI disks. When Windows NT detects a bad sector on a disk, it attempts to move the data to a good sector, rebuilding it from the redundant copy if necessary. The disk driver then requests the hardware to map out the bad sector. If the process is not successful, the administrator is notified through a system event message.

Windows NT Architecture

Architecturally, Windows NT was designed using an object model. This means that most of the services and resources provided by the operating system files, such as directories and printers, are viewed as objects. By using an object model, Windows NT provides a modular approach to the operating system. Therefore, the various components operate independently, allowing them to be modified without breaking the entire system. Windows NT moves beyond the 640 kilobyte limit of MS-DOS, utilizing a flat 32-bit memory model. Each application is allocated a 4 gigabyte virtual memory space.

Networking is built into Windows NT, including both client and server capabilities.

This allows any NT computer to participate in a network in either capacity. Windows NT networking architecture includes both peer-to-peer and server based networking. With no additional software required, Windows NT can interoperate with the following networks:

- Microsoft networks
- Novell NetWare
- TCP/IP, including UNIX hosts
- Apple Macintosh Appletalk (NT Server only)
- Remote access clients

As with other portions of the operating system, Windows NT networking architecture is based on a modular component design. This modular approach enables modification of the various components with minimal effect on the other components. Windows NT includes two important interfaces in its networking model:

1. NDIS 3.0 Interface (NDIS)
2. Transport Driver Interface (TDI)

These interfaces isolate the layers of network component software so that each adjacent component can be written to a common programming interface. This allows the various components to be portable and interchangeable. For example, network card drivers are written to the NDIS 3.0 Interface. Therefore, each network card driver does not need specific programming code to access the various transport protocols. This access is handled by the NDIS 3.0 Interface and the related transport protocol drivers. Windows NT supports two basic networking models: peer-to-peer and domain.

Peer-to-peer networking is also referred to as the workgroup model and typically is used in relatively small networking environments. Peer-to-peer networking is the only model supported by Windows NT Workstation. In this networking model, each computer in the peer workgroup maintains its own user accounts and security database.

The domain model is essentially a logical grouping of server resources and users. The primary purpose of the domain is to centralize management of network resources and users. The domain, for practical purposes, forms an administrative unit that allows centralized management of network resources and users.

The domain maintains a single database of all user and group accounts and security information. Only one account needs to be maintained for each user in the domain. When users log on to the domain, the logon information is validated against the user database, ensuring that users are authorized to access network resources.

Domains also allow users to browse the network for available resources. Network resources are grouped by domain. Therefore, users see resources only of selected domains, rather than all the servers and printers on the network.

Windows NT Registry

The registry utilized by Windows NT is a database that stores all the configuration information for the operating system. Microsoft supplies the registry editor program (regedt32.exe) with Windows NT to allow users to directly edit the registry. Most registry modifications are made through the various control panel applications supplied with Windows NT. For example, the Network control panel makes the changes to the registry needed to configure the various networking components found in Windows NT. It must be cautioned that using the registry editor improperly can cause severe system-wide problems, with the only remedy being to reinstall Windows NT. Microsoft advises that you use this tool at your own risk.

Windows NT Market Applications

Windows NT Workstation is primarily marketed as a high-end desktop workstation. Its major competition, both technically and in the market, is IBM's OS/2 operating system. Because of its Microsoft heritage, it integrates well into the Microsoft Windows environment. With its 3GL graphics capability, combined with higher-end Pentium and RISC processors, it also challenges UNIX based workstations. Its support for peer-to-peer network connections gives it additional appeal in the small network environment, where the need for a dedicated network server may not exist. Its support for multiple hardware platforms allows NT users a variety of hardware platforms, and support for evolving new hardware platforms is virtually assured.

Windows NT Server, as the name implies, is aimed at the enterprise server market, as a file, print, and application server. While its primary competition is Novell NetWare, the true battleground is the upgrade path from Novell NetWare 3.x. In this market, Microsoft and Novell are both trying to be the "path of choice" for the many NetWare 3.x networks already in place.

Microsoft also offers an integrated package of tools for Windows NT Server, called BackOffice. BackOffice, which is sold and licensed separately from Windows NT Server, adds the following components to NT Server

- SQL Server—SQL Database
- SNA Server—IBM mainframe connectivity
- Systems Management Server—Hardware/software maintenance
- Microsoft Mail Server—Electronic mail system

With these additional tools, Microsoft is attempting to give enterprise network administrators a one-stop-shopping approach to their network software and maintenance needs.

Windows NT and NetWare

Windows NT was designed to interoperate with Novell NetWare. In an overall view, there are three levels of NetWare interoperability:

1. Basic Connectivity
2. Gateway Services for NetWare
3. File and Print Services for NetWare

Basic NetWare Connectivity

Basic NetWare connectivity requires installation of the following components on either an NT Workstation or Server computer: NWLink and Client Services for NetWare

NWLink is Microsoft's 32-bit IPX/SPX compatible protocol. It supports Novell NetBIOS, Windows Sockets, and RPC (on top of SPX). NWLink is routable, allowing Windows NT systems to communicate with other IPX routers. Use of NWLink also allows NetWare clients to run the server portion of client/server applications, such as SQL Server, on the Windows NT computer.

Client Services for NetWare (CSNW) is a 32-bit native NetWare redirector for Windows NT. It includes both a service and a device driver. CSNW includes the following functions:

- NT user access to NetWare file and print servers
- NetWare application support
- Support for NCP, Burst Mode, and LIP protocols
- Long file name support, when NetWare is running the OS/2 name space

See Figure 5-23 for a list of connectivity options for Windows NT and NetWare clients.

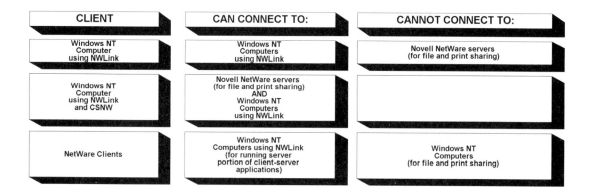

Figure 5-23. NT and NetWare Interoperability

Gateway Services for NetWare

Gateway Services for NetWare (GSNW) is available only on Windows NT Server. By installing this service, the NT Server computer acts as a gateway for its clients, allowing them to use resources available on the NetWare Network. Unlike the Client Services for NetWare, Microsoft network clients do not need to run the NWLink protocol to access the NetWare server file and print resources (See Figure 5-24).

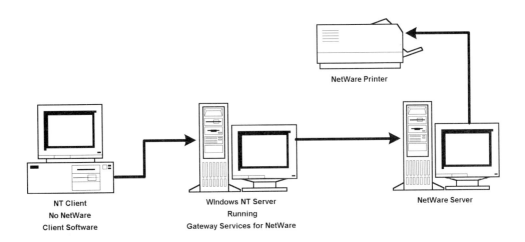

Figure 5-24. NT NetWare Gateway Services

As with basic NetWare connectivity, the NWLink protocol must be installed. Gateway Services for NetWare is compatible with file and print services on NetWare 2.x and 3.x servers and on NetWare 4.x servers running bindery emulation. NetWare 4.x NetWare Directory Services is not supported.

GSNW is intended as an occasional use router, because all NT client users are accessing the NetWare network through one NetWare connection. Excessive access to the NetWare network through the NT gateway could cause performance to suffer.

File and Print Services for NetWare

File and Print Services for NetWare (FPNW) is an extra cost service that allows NetWare clients access to a Windows NT Server. Essentially, FPNW allows a Windows NT Server to act like a NetWare 3.12 file and print server. NetWare clients can then access the file and print resources of the Windows NT Server. FPNW is a critical tool that allows the integration of a Windows NT Server into an existing NetWare 3.12 network.

Specifically, FPNW supports the following functions:

- File access and management using NetWare tools
- Creation and management of user accounts
- Printing and print queue manipulation
- Remote administration
- Secured logins

Windows NT and the Internet

Windows NT supports the TCP/IP protocol, and is making a strong showing as an Internet server. Windows NT Workstation and Server both ship with a variety of TCP/IP support, including:

- TCP/IP protocol and utilities
- FTP Server Service

The Windows NT Resource Kit includes the following additional support:

- World Wide Web, Gopher, and WAIS Server Services
- Domain Name System (DNS)
- Windows Internet Name Service (WINS)

Windows NT Server includes Dynamic Host Configuration Protocol (DHCP), which allows the NT Server to dynamically allocate IP addresses to client workstations as needed.

TCP/IP Protocols and Utilities

Windows NT includes the core TCP/IP protocols, such as UDP, ARP, and ICMP. It also includes basic TCP/IP Utilities, such as Finger, FTP, RCP, REXEC, RSH, Telnet, and TFTP. SNMP services and TCP/IP printing are also supported.

Using Remote Access Service, Windows NT can connect to the Internet via a SLIP or PPP connection. Windows NT can also act as an Internet router for small networks, and can route local TCP/IP packets to the Internet via the SLIP or PPP modem connection.

Internet Services

With Windows NT, it is fairly simple and inexpensive to configure an NT Workstation or Server to operate as an Internet server. FTP Server Service is included with the operating system. With the addition of the Windows NT Resource Kit (or via FTP), Windows NT can also function as a World Wide Web, Gopher, and WAIS server.

SUMMARY

As you review the histories of architectures of the PC LAN vendors, an interesting pattern emerges. These vendors are not unlike the major computer vendors, including IBM and Digital, who began their businesses by developing proprietary systems and who, over time, have bowed to customer demands for standardization.

As PC LANs have grown in importance within organizations—both commercial and government—customer demands to integrate them into the enterprise level network have increased in intensity. And the vendors have responded.

For customers who need mainframe connectivity, PC LAN vendors have provided SNA gateways. Customers who need DEC connectivity most often purchase DEC PCs and network them through DECnet. But for the few who are not pure DEC, PC LAN vendors have provided asynchronous communications gateways. And, as UNIX-based PCs have made their way into the world of UNIX-based workstations, products such as PC NFS have provided connectivity.

But does connectivity mean interoperability? In many companies today, we find very complex, multivendor, multiprotocol environments. Users want to share files, transfer data, and send messages across multiple platforms. The trend among mainframe, minicomputer, workstation, PC, and LAN vendors is to support this demand through TCP/IP, with most vendors providing Ethernet and Token Ring connectivity solutions at the Physical and Data Link Layers.

With TCP/IP, users can share and transfer files, exchange mail messages, access the host through terminal emulation, and access print and storage resources available throughout the network.

Most PC LAN vendors are working to make their LAN Network Operating System (NOS) a Enterprise or Corporate Network Operating System (CNOS). This will require seamless internetworking with support of a variety of protocols, hosts, and applications—with true peer-to-peer networking. All of this must be accomplished independent of transport mechanisms, protocols, and operating systems. Their goal is, in fact, interoperability rather than mere connectivity, with support of true peer-to-peer, client/server applications.

PC LANs have evolved from purely local networks, with PCs simply sharing printers and hard disks, to networks that take advantage of resources throughout the enterprise; the client/server model has made this possible. The open architecture of the client/server model supports multiple protocols, enabling organizations to run existing DOS, OS/2, DECnet, UNIX, Windows, and Macintosh applications, while supporting expansion into advanced distributed computing companywide and, often, worldwide.

In the client/server model, the server and its clients share computing activities. The server performs database access and intensive computing tasks, and the client performs the display and user interface tasks for the requested calculations. This model distributes the computing load throughout the network and takes advantage of the power available on the desktop.

Network Operating System (NOS) software resides at the equivalent of the Session and Presentation Layers of the OSI model, with some Application Layer functions often included. The NOS evolved primarily to give PC users access to other users' printers and disks and to files stored on a PC LAN server. The integration of the LAN into the larger corporate networking environment has expanded the role of the NOS. Now the NOS is responsible for providing information access to users of multiple protocols, operating systems, and hardware platforms.

The primary vendors in the PC LAN market are Novell, with its very powerful and popular NetWare and Portable NetWare products; Microsoft with Windows NT; Banyan, the leader in wide-area networking; and UNIX based LANs with VINES. Apple, with Phase II of AppleTalk, has bowed to industry pressure to adopt an open systems approach to networking Macintoshes, especially as the Mac has infiltrated the corporate environment. Similarly, IBM has responded to customer demands that PCs have a window into the enterprise-level SNA network with OS/2 LAN server.

ADDITIONAL INFORMATION ON THE CD-ROM

- Additional information on terms and subjects in this chapter
- 3-COM NOS
- TCP Traces
- LAN Manager NOS
- RIP Protocol
- ARP Traces
- IPX/SPX Traces
- NetBIOS Traces
- AppleTalk Traces
- SPX Traces
- IP Traces
- NCP Traces
- Echo Protocol
- Matrix of NetWare Products
- UDP Traces
- VIPC Traces
- AppleTalk Protocol formats
- IPX
- SMB Traces

6

TCP/IP ARCHITECTURE

INTRODUCTION TO TCP/IP

The acronym TCP/IP stands for the names of two protocols developed for the ARPANET. However, it is often used to denote more than just the two protocols themselves. The TCP/IP "world" includes three components: The ARPANET and the networks to which it is linked, when taken collectively, are called the Internet. The entire suite of protocols, software, and applications are standard parts of most UNIX-based systems, including Ethernet and applications such as FTP (See Figure 6-1). The actual protocols are the Transmission Control Protocol and the Internet Protocol. This chapter focuses on the protocols and on the applications built upon them (which logically form the application layer, though such a layer is not explicitly defined). You studied the Ethernet protocol in Chapter 4; you will study the Internet and internetworking aspects of TCP/IP in Chapters 9 and 10. This chapter will provide an understanding of TCP/IP "autonomous systems," that is, TCP/IP-based networks that are isolated—the kind of networks you might find in a corporation that has purchased many workstations to use internally for CAD/CAM and other engineering and manufacturing applications. In focusing on autonomous systems, this chapter will include information on the interconnection of "subnets" to create TCP/IP networks. But the broader topic of internetworking, that is, the interconnection of autonomous systems, is reserved for Chapter 9.

CHAPTER OBJECTIVES

- Identify the TCP/IP layers, their components, and their functions

- Name the OSI layer that corresponds to each TCP/IP layer

- Identify the components of a class B Internet address in the dotted decimal form

- Give the general form of a hierarchical Internet name

- Explain how LAN addresses are determined

- Give the purpose of the ARP cache and state what is stored in it

- Tell how a gateway node differs from an ordinary node

- Characterize the delivery service provided by IP and TCP

- Identify the reasons that an application might use UDP

- Identify the services that six important TCP/IP applications provide

Figure 6-1 shows the relationship between TCP/IP and the OSI model. Remember that TCP/IP predates the OSI model by about a decade, so that the layers of the two architectures do not cor-

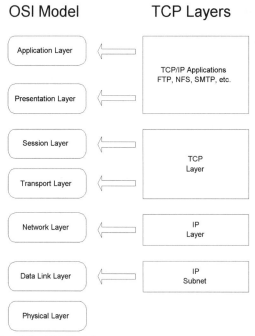

Figure 6-1. TCP/IP and the OSI Model

respond very well to one another. TCP/IP has only two explicitly defined layers, the Transmission Control Protocol layer, TCP, and the Internet Protocol layer, IP. The application layer, which is very rich, is not explicitly defined. Tasks which are taken care of by the OSI Session, Presentation, and Application Layers, such as the translation of data formats between different processors and the managing of dialogs, are left to the individual TCP/IP applications for resolution.

TCP/IP originated long before personal computers and workstations, in the days when network computers were all multiuser minicomputers and mainframes with many dumb terminals attached. For this reason, some TCP/IP terminology is different from terminology that has been used in this book. Because every network node "hosts" many users on dumb terminals, TCP/IP uses the term host in the same sense as node has been used in this course. To remain consistent with the rest of the book, we will continue to use the term node in this chapter. If you encounter the term host in other materials written about TCP/IP, you can mentally substitute the term node. Figure 6-2 shows the relationship between networks, subnets, and nodes.

The extent to which networks would proliferate could not have been foreseen by the early architects of ARPANET. LANs had not been invented at that time. The Internet addressing

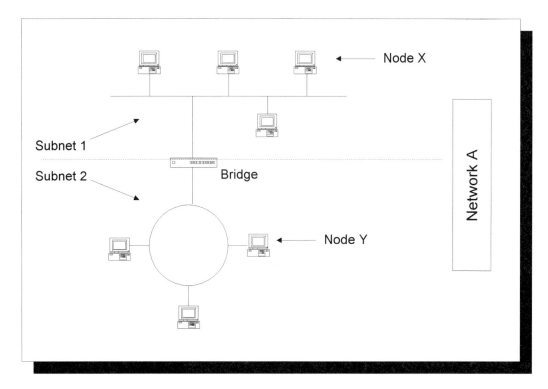

Figure 6-2. TCP/IP and Subnets

scheme that they devised, consisting of only a network address and a node address, didn't allow for the many LANs that exist today. To solve this problem, the concept of a subnet was born. A subnet is essentially a single LAN that is part of an Internet network. You will see how the Internet addressing scheme handles subnets.

A network can be a collection of "hosts," that is, independent nodes connected by point-to-point links; it can be a single LAN which has its own network ID; it can be a collection of LAN subnets linked together by nodes which are connected to more than one LAN; or it can be any combination of these possibilities.

Figure 6-3 shows a more detailed picture of the TCP/IP layers and includes some components which have already been introduced and some which are new. We have again shown the correspondence to the OSI layers. This chapter will explain the services and protocols of these layers in the context of an autonomous network. We will describe these layers briefly here, and then study each in more depth.

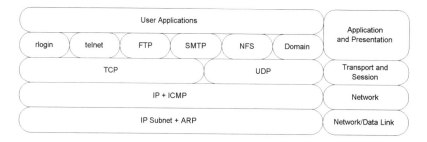

Figure 6-3. TCP/IP Layers

TCP/IP Layer Overview

IP Subnet + Address Resolution Protocol (ARP)

The IP subnet functions of IP, also called the Network Interface, use a variety of lower level protocols, such as the IEEE 802.2 Logical Link Control (LLC) that you learned about in the previous chapter, to deliver datagrams. The ARP is used to associate an Internet address (explained below) with a LAN hardware address (NIC address).

Internet Protocol (IP) and the Internet Control Message Protocol (ICMP)

The Internet Protocol layer is responsible for connectionless, unreliable, but best effort, packet delivery service to the upper layers. The protocol is embodied in an Internet datagram. ICMP, an integral part of IP, delivers messages used by IP to control the network.

Transmission Control Protocol (TCP)

The TCP layer is responsible for providing connection-oriented, end-to-end, error-free message delivery. Since it uses the IP layer's unreliable delivery service, its major responsibilities are error correction and connection management.

User Datagram Protocol (UDP)

An alternative to TCP, UDP provides a simple, connectionless, but error free, datagram delivery service for certain specialized application services which do not require the full services of TCP.

TCP/IP applications and application services include the following:

- **rlogin.** Allows remote login from one TCP/IP system to another.

- **Simple Mail Transfer Protocol (SMTP).** It provides electronic messaging for mail service and for Network News.

- **TELNET.** The TCP/IP standard protocol for connection of remote terminals.

- **File Transfer Protocol (FTP).** It provides for the transfer of files between systems. FTP uses TELNET to establish a control connection.

- **Domain Name Protocol (DOMAIN).** Converts a hierarchical Internet domain name. For example, Sales.mycompany.com to an Internet address, such as, 144.49.4.70.

- **Network File System (NFS).** Developed by Sun Microsystems, NFS is the de facto standard for file sharing.

- **IP Subnet Layer.** Also called the Network Interface Layer. The nature of the low-level IP Subnet Layer depends on the communication link to which the node connects. If the node connects to a LAN, then the IP Subnet Layer will consist of NFS routines based on a protocol such as the IEEE 802.2 LLC protocol. If the node connects to a remote subsystem, then the IP Subnet Layer will consist of a complex subsystem which uses a data link protocol such as HDLC. In either event, the IP Subnet Layer accepts IP datagrams and transmits them over the specified link, using ARP (explained below) to determine the LAN address as necessary.

TCP/IP Naming and Addressing

The subject of the naming and addressing in TCP/IP can be somewhat confusing, unless several basic conventions and concepts are understood. The Central Authority, who is responsible for naming all networks which will be interconnected into an Internet. For small internets, this might also be the Network Administrator for the networks. For the Internet, the Network Information Center (NIC) is responsible for assigning values and attributes for the overall Internet. Listed below are important terms and concepts:

- **Network Administrator.** Responsible for naming within a network or for several networks at a site.

- **System Adminstrator.** Responsible for naming nodes in a subnet. Often the same person as the Network Administrator. Names or addresses must be assigned at several levels.

- **Network ID.** A network has an address which is a number assigned by the Central Authority and guaranteed to be unique for networks on the Internet.

- **Subnet ID.** A subnetwork has an address which is a number, unique to the network, assigned by the Network Administrator.

- **Node ID.** Each node in a network has an address which is a number within the subnet, assigned by the System Administrator.

- **LAN ID.** Each node in a LAN has a subnet address hard-wired into the machine, but not the same number as its network address. The LAN ID would typically be referred to in terms of the type of LAN. For example, for an Ethernet LAN, the LAN ID would be referred to as the node's Ethernet address.

In order to make naming easier for users, TCP/IP supports a hierarchical naming system to allow names to be assigned to machines and networks. For example, this system can be used to address e-mail messages.

- **Node Name.** Each node in a network has an alphanumeric name; for example, sales.

- **Domain Name.** Given to a network or group of networks. The Central Authority can establish conventions for domain names. The general form of a domain name is alphanumeric characters separated by periods, in the form nodename.xxx.yyy.zzz. The rightmost portion of a domain name (zzz) is the name of the domain itself; the names relate hierarchically from right to left, with the node name at the far left. A domain name might look like this: sales.mycompany.com.

- **Internet Conventions.** For the Internet, these conventions have been adopted for domain names.

The Internet defines several high-level domains (the rightmost name). Commercial sites, such as most companies, belong to the com domain. Each site on the Internet has a unique name which is registered with the Central Authority. A site name can cover more than one network ID. Finally, within a site, networks and nodes can be further grouped and named.

- **User Name.** Each person who uses a node has at least one alphanumeric login ID. For example, John Doe at Company might have a user name of johnd, and use a VAX work-station named engr which has a LAN ID of 123456 . His node might have an Internet address of 129.49.4.33 which gives the Network ID, the Subnet ID, and the Node ID. John can be addressed as johnd@l29.49.4.33 or johnd@engr.company.com.

Let's look at each of these and how they are related.

Internet Addressing of Networks, Subnets, and Nodes.

An example of an Internet address is 144.49.4 .42. Internet addresses are customarily written in that form—four decimal digits separated by periods, called dotted decimal notation. An Internet address is actually a 32-bit binary number (a four-byte word). It is written as four decimal integers to make it easier for humans to handle. Each of the four integers represents a byte of the address, so the largest value that can ever occur is 255, and the theoretical range of possible addresses is 0.0.0.0 to 255.255.255.255 (hex '00000000' to hex 'FFFFFFFF'). (Remember, in hexidecimal notation, an "F" is equal to four ones and a "0" is equal to four zeros.)

Internet address 128.33.68.2 is equivalent to binary 1000 0000 0010 0001 0100 0100 0000 0010, which is equivalent to hexadecimal 80 21 44 02.

An Internet address incorporates a network address, a subnet address if there are subnets, and a node address. These three pieces of information can be combined within the single binary word in several ways. The network address can take one, two, or three bytes, leaving three, two or one bytes for the combined subnet/node address. To further complicate things, the byte or bytes used for the subnet/node address can be divided arbitrarily, with certain bits for the subnet address and certain bits for the node address, as explained below. Fortunately, as a practical matter in a given installation, only one of the many possible schemes of addressing is used.

Every node on the network must know how to tell which bits in the Internet address correspond to their physical network, or subnet. This is accomplished through a "subnet mask" which is set through the software in each node. If the subnet mask is incorrectly set by the user or System Administrator, the node will not be able to recognize its address in messages on the LAN and will not be able to communicate. The subnet mask, which must be consistent throughout the network, is a 32-bit hexadecimal word which "masks out" the node address.

The Internet addressing scheme used by TCP/IP networks was devised to permit addressing of nodes anywhere on the Internet network. You will learn more about the Internet and addressing for the Internet itself in a later chapter. If the network in question is attached or is planned to ever be attached to the Internet, then the network address portion of the Internet address is assigned by the Central Authorities, thus guaranteeing that every network on the Internet is uniquely addressed. Central Authorities of TCP/IP networks which do not attach to the Internet are free to assign network addresses by whatever scheme appeals to them, but the 32-bit address and dotted decimal notation are required by the TCP/IP software.

Nodes on an Ethernet or Token Ring LAN have within them communication circuits which have "hard-wired" 32-bit network addresses. (Actually, the addresses are usually stored in read-only memory and can be changed). The addresses incorporate a unique number assigned to each manufacturer, and the manufacturer assigns unique serial numbers to each device. As you have seen, the Internet address contains both a network address and a node address. But the node address in the Internet address is not the same number as the hardware LAN address. The Internet node address is unique only for the network and is assigned by the Network Administrator.

A TCP/IP node always knows both its LAN address and its Internet address. When the node needs to transmit an IP packet, it can tell whether the target node is on the same LAN by looking at the network address to see if it is the same as its own. When the target is on the same LAN, it can transmit the packet directly to the target. It must, of course, know the LAN address of the target to do so. In other words, there must be a mechanism to associate the Internet address with the LAN address. That's where ARP comes in. ARP is a low-level protocol that permits TCP/IP nodes to use Internet addresses transparently. ARP works like this: Each node keeps a cache of addresses for nodes on its own LAN. The table is empty when the node is booted. When an address translation is required and the translation is not in the cache, the ARP routine broadcast a special packet, called an ARP packet, to every other node on the LAN. The packet essentially asks, "Does anyone know the LAN address for ww. xx. yy. zz?" Each node on the LAN processes the message by comparing the transmitted Internet address to its own. The node with a matching address replies, broadcasting an ARP packet saying, "I'm ww.xx.yy. zz and my LAN address is nnnnnn." This response gets stored in the cache on every node on the LAN, so very quickly each node will have a complete cache. The ARP messages will not have to be sent again until a node is booted or a new node is added to the LAN.

IP is responsible for providing a datagram delivery service that is both connectionless and stateless. It must be connectionless because IP has no concept of a circuit or a stream of data. It must be stateless because IP makes no connection whatsoever between two datagrams which happen to be addressed to the same destination.

IP makes no attempt at error correction or recovery. No checks for correctness of data are performed at this level. If a message is garbled due to a transmission error, IP will simply pass the garbled information up to the next layer. If it encounters an error condition which it can detect, but from which it cannot recover, IP will simply throw the datagram away.

Although it is unreliable, it will not throw data away capriciously. For example, if it encounters an error during transmission, it will not discard the datagram until it has made a thorough attempt to recover by retrying the transmission.

Network Layer Concepts

In order to send a message from one node to another, IP starts by accepting requests from the transport layer to send a packet, along with the Internet address of the machine to which the packet is directed. Then if the length of the packet is greater than that which can be accommodated by the physical link protocol, IP divides the packet into fragments. IP encapsulates each fragment by adding an IP header in which it stores the Internet address of the source and the destination machines and several other control fields. If the destination node is on the same subnet (if it has the same network and subnet ID), IP checks the address cache or, if necessary, uses ARP to determine the subnet address of the destination node. If the destination node is on a different network or subnet (has a different network or subnet ID), IP sends the fragment to a router or gateway which provides connectivity to the distant network. Having completed these steps, IP passes the fragments to the Link Layer for transmission.

At the receiving end, IP removes the IP header from each fragment that it receives from the Link Layer. Using the fragment number in the header, IP reassembles the fragments into packets. The IP layer processes on the sending and receiving nodes will cooperate using the Internet Protocol and the ICMP sublayer.

Subnets are connected to one another via gateways (routers). A gateway can be a special device which performs the interconnection function only, but far more often it is also a node on the network. A gateway node resides on two or more LAN subnets. It is responsible for routing datagrams from nodes on one subnet directly to nodes on others to which it is attached, and indirectly to other subnets.

A node becomes a gateway by adding a second LAN connection (typically by adding a board to the chassis). Since each LAN circuit has its own unique LAN address, the gateway node has two LAN addresses. It also has two subnet addresses, since it resides on two different subnets. If the target node is on another subnet to which the gateway is attached, it simply transmits the fragment to that node. If the target node is not on a subnet to which the gateway is attached, then there must be a route to the target through another gateway. IP inspects its routing tables to determine which gateway is next on the route to the target and then transmits the fragment accordingly.

This section has explained routing messages between subnets. Routing messages between networks will be addressed in Chapter 9.

Transport Layer Concepts

Although TCP uses IP as the delivery system in the vast majority of cases, there are exceptions. For example, TCP can use the LAN directly on a single subnet. This discussion does not assume that IP is the underlying delivery system, although that is usually the case. The job of TCP is to convert an underlying delivery system, which is assumed to be connectionless and might be unreliable, to a communication channel that has these qualities:

- **End-to-End**. While the underlying delivery protocol layer, like the Network Layer of OSI, sees only the nodes to which it is connected, TCP sees the whole network—in fact, the whole internetwork.

- **Connection-Oriented Stream Transport**. Like the OSI Transport Layer, TCP establishes connections and guarantees that messages which might have been divided into packets (and divided further into fragments) are fully reconstructed and delivered in the sequence in which they were transmitted.

- **Full Duplex**. A TCP connection provides two channels, one in each direction. An application can use the channels to overlap data flow, or it can use one channel for data and the other for control information.

- **Error Free**. TCP takes care of detecting damaged datagrams, replacing damaged or lost datagrams, and discarding duplicate datagrams in order to guarantee that valid data are delivered to the destination. Transmission errors are detected by the use of checksums. Lost and duplicate datagrams are detected by checking the necessary information in the protocol headers. All datagrams are acknowledged, and missing or lost datagrams are retransmitted.

- **TCP is Independent of IP**. Although in most cases IP will be the underlying protocol layer, that is not necessarily always true. For example, Ethernet alone can provide the datagram service if there are no interconnected subnets.

- **User Datagram Protocol (UDP)**. Applications cannot send IP datagrams themselves because the IP protocol contains no provisions for a datagram to be addressed to a process within a node. In other words, if an application could send a datagram, the IP process in the target node would have no way of knowing what to do with it when it got there. Although applications can use TCP, in some cases TCP provides more services than are required, therby unnecessarily increasing overhead. UDP encapsulates a user message with a header that provides the information required for a process within one node to communicate with a peer process within another node. So the unreliable, connectionless datagram service of IP is available to applications which do not require the additional services (and overhead) of TCP.

TCP/IP Applications and Application Services

The TCP Layers diagram (Figure 6-3) can be used for reference with the following description of Applications and Applications Services.

rlogin

rlogin stands for remote login. Suppose that you are logged in on your own workstation, but need to run a complicated inquiry on a database that resides on another workstation. You could, of course, transfer the file to your system via FTP, but if the file is very large, that could take hours. You could mount it via NFS, but response time could be very slow because your database software would access the data records across the network.

Assuming you have a login ID established on the other system, you can simply rlogin to the other system and execute commands as if you were physically connected to and logged into that workstation. Therefore the database activity would be local to the database file system, and only the input statements and the displayed or printed results would have to be transmitted across the network. Because rlogin is a UNIX-to-UNIX facility, the other system understands the environment you are running in. For example, you could direct the output of an inquiry command to be stored in a file on your own system rather than on the remote system.

SMTP

Unlike other communications protocols, which use binary codes in structured fields, SMTP uses plain English headers. SMTP defines a protocol and a set of processes that use the protocol to transfer e-mail messages between users' mailboxes. It does not define the programs used to store and retrieve mail messages. In fact, although a basic mail "reader" program is included with TCP/IP on virtually every operating system, many different mail readers have been developed. These include programs based on graphical user interfaces such as Motif and OPEN LOOK. But they all use the SMTP service routines to send and receive.

TELNET

TELNET is a virtual terminal protocol essentially similar to that mentioned in the discussion of the OSI Application Layer in Chapter 3. Virtually all nongraphical applications written for the TCP/IP environment use the TELNET protocol for input and output to the user.

Unlike rlogin, TELNET allows you to behave like a terminal accessing a remote node, so the UNIX system will not, for example, direct its output to a file on your system. That is the disadvantage of TELNET, compared to rlogin. TELNET has the advantage of allowing communications between UNIX TCP/IP systems and many other types, while rlogin is only UNIX-to-UNIX communication.

FTP

FTP actually creates two independent channels between the nodes: a control channel and a data channel using both TCP and UDP. The control channel uses TELNET for communication between the user and the node that is remote to the user, for example, to specify the names of files that are to be transferred. Data are transferred through a TCP stream on the other channel.

DOMAIN (Domain Name Server or DNS)

As you read in an earlier section of this chapter, a user can communicate with another user by specifying a domain name, such as johnd@engr.company.com. TCP and IP require Internet addresses for messages, so one must be translated to the other. That is the job of the name server—given a domain name, they will return an Internet address (but not the reverse). Recall that domain names are hierarchical. There is a corresponding hierarchy of name servers. Given the domain name kdr@mcdata.com, DNS will resolve this from right to left. DNS will first locate a server on the Internet that knows all of the locations for companies. That DNS will then locate the company McDATA and ask for the address of kdr.

NFS

NFS lets a user mount a file that resides on another node and access it as if it resided on the user node. Although it can take considerably longer to transfer data through NFS than through the local file system (because data transfer across the network is significantly slower than data transfer across a disk channel), the alternative is to make a local copy of the data (using FTP, for example). Making the local copy requires considerable initial overhead and gives rise to data integrity problems because multiple copies of the data exist.

The following steps allow NFS file sharing:

1. The node upon which the data reside (the server) must "export" the directory in which the file is listed by putting the directory name in a special system file, /etc/exports. The files in an exported directory can be available to all nodes or to specific nodes identified by the System Administrator. The user controls exported directories. Normal file access privilege procedures apply to the exported files. For example, they can be marked "read only."

2. The node desiring to access the data (the client) "mounts" the exported directory at a specified point within the client file system. It appears to the user on the client system that the exported directory becomes a new subdirectory of the user's own file system.

NFS extends "file locking" to remote files. Programs which access local files that are shared with other processes on the node can "lock" the whole file or some portion of it during updates. This prevents the problem of two programs updating the file simultaneously, which will result in the loss of one of the updates or will cause inconsistencies in the data in the file. Note that access privileges are by node name, not by user name, so any user on a node that has mounted a remote file has access to that file.

Internet Addresses

Each host on a TCP/IP Internet is assigned a unique 32-bit Internet address that is used in all communication with that host. Conceptually, each ID is a pair of addresses, the Netid and the Hostid. Netid identifies a network and Host ID identifies a host on that network. The three most common IP addresses formats (class A, B, and C) are shown in Figure 6-4. (There are actually five classes of IP addresses used for different size networks). Gateways base routing decisions on the NETID portion of the address. Since some machines (such as routers) have two connections to a network, IP addresses specify connection to a network, not an individual machine (HOSTID 0). A broadcast transmission uses a Host ID of all 1s (hex FF).

Figure 6-4. Internet Address Formats

Dotted Decimal Notation

IP addresses are normally seen as four decimal integers separated by decimal points where each integer gives the value of one octet of the IP address. Therefore the address is:

10000000 00001011 00000100 00011110 = 128.11.4.30

Loopback address = 127.0.0.0 is used to test the protocol software of the computer and is not sent across the network.

Class A addresses are used on networks that have a large number of IP hosts (16,777,216 or less). The first byte of a Class A address begins with a number from 1-127. Class B addresses begin with a number from 128 to 191, and Class C addresses have three bytes which represent the network portion of the address. Class C addresses can identify from 1-256 unique hosts on a single network.

Disadvantages to Addressing Scheme

1. If a host moves from one network to another its address must change.
2. The path taken by packets traveling to a host may be different if multiple addresses refer to the same host.

TCP/IP Protocols

ARP and RARP

ARP (Address Resolution Protocol) maps Internet addresses to physical addresses. Machines located on physical networks can only communicate if they know each other's physical network address. If machine A wants to send a message to machine B and only knows the IP address, there must be a way to map the IP address to the physical address. This will allow higher level programs to work with Internet addresses.

How does this work with Ethernet? Ethernet contains a 48-bit address which is "burned in" to the actual interface board in the machine. If this board fails, the replacement board will contain a new physical address. ARP allows a machine to find the physical address of a target host on the same physical network, given only the IP address. An ARP message is encapsulated in the frame of the physical network.

The ARP protocol format is designed to be used with a variety of network technologies. See Figure 6-5 for an example of ARP protocol format. The fields of this diagram are described below:

- Hardware Type: specifies interface, i.e. 1 for Ethernet
- Protocol Type: specifies high level protocol address, i.e. IP = 0800
- Hlen and Plen: specifies length of hardware address and length of protocol address
- Operation: ARP/RARP request or response
- Sender HA/IP: sender hardware and IP address
- Target HA/IP: sender target IP address (ARP) or hardware address (RARP)

32 BITS

HARDWARE TYPE		PROTOCOL TYPE
HLEN	PLEN	OPERATION
SENDER HA		
SENDER HA		SENDER IP
SENDER IP		TARGET HA
TARGET HA		
TARGET IP		

Figure 6-5. Address Resolution Protocol Format

RARP

Reverse Address Resolution Protocol is used by diskless machines to find out their IP address. The machine broadcasts a message on the network to a server, which responds with the IP address of the diskless machine.

IP Datagram Format

Figure 6-6 shows an example of the IP Datagram Format. The descriptions of each of the fields in this diagram are listed below:

- VERS: This 4-bit field in the datagram contains the version of the IP protocol that was used to create the datagram. It is used to make sure that the sender, receiver, and gateways in between agree on the format of the datagram.
- HLEN: This 4-bit field is the header length field and gives the length in 32-bit words. The most common header, which contains no padding and no IP options, measures twenty octets and has a header length field of five.

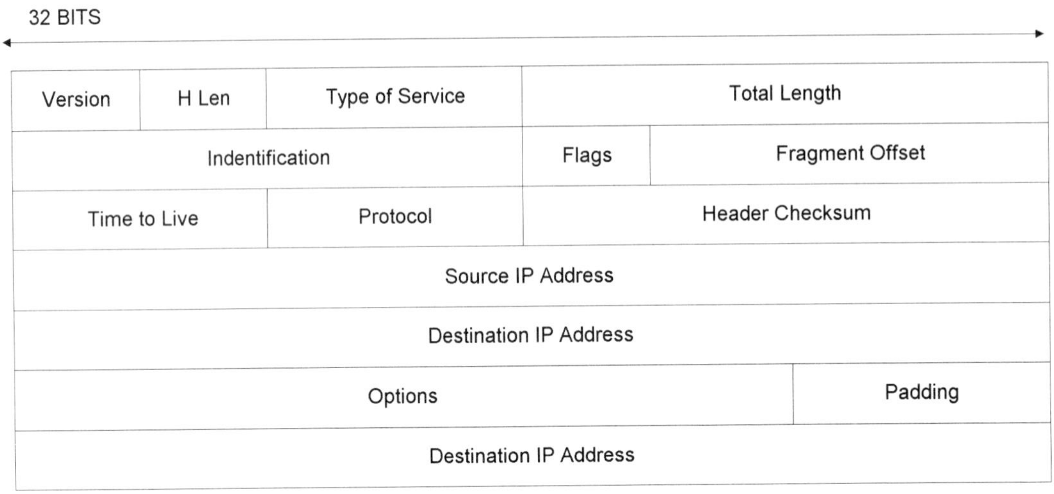

32 BITS

Version	H Len	Type of Service	Total Length	
Indentification			Flags	Fragment Offset
Time to Live		Protocol	Header Checksum	
Source IP Address				
Destination IP Address				
Options				Padding
Destination IP Address				

Figure 6-6. Internet Protocol Format

- TYPE Of SERVICE: Specifies how the datagram should be handled.
- TOTAL LENGTH: Gives length of IP datagram in octets, including the header and the data. The maximum size is 65,535 octets.
- IDENTIFICATION, FLAGS, AND FRAGMENT OFFSET: Control the fragmentation and reassembly of datagrams. Most datagrams are encapsulated into multiple frames of a smaller size (Ethernet max MTU is 1500 bytes).
- IDENTIFICATION: Each unique datagram must have a unique identification number to identify the datagram.
- FRAGMENT OFFSET: Specifies the offset in the original datagram of the data being carried in the fragment, measured in units of 8 bytes starting at 0.
- FLAGS: Controls fragmentation.
- TIME TO LIVE: Specifies how long (in seconds) the datagram is allowed to remain in the Internet. Gateways and hosts decrement this field and remove the datagram when its time expires. This field is also referred to as the hop count when the value is decremented at every routing device.
- PROTOCOL: Specifies which high-level protocol was used to create the message being carried in the DATA area of a datagram.
- SOURCE/DESTINATION ADDRESS: Specifies the IP address of the sender and expected recipient. They do not change.
- IP OPTIONS: Variable length field used for various options such as recording the route taken, specifying the route to be taken, and time stamping.

Routing IP Datagrams

Direct vs. Indirect Routing—direct routing refers to routing between two machines on a single physical network which does not involve gateways. The sender encapsulates the datagram in a physical frame, binds the destination address to the physical address, and sends the frame directly to the destination. Indirect routing is more difficult because the sender must identify a gateway to which the datagram can be sent. Datagrams pass from one gateway to another until they reach a gateway that can deliver the datagram directly. Internet routing tables are generally used to route datagrams using the network portion of the datagram address.

Transport Control Protocol Format

Segments are the unit of exchange between two machines using TCP software. These segments are exchanged to establish connections, transfer data, send acknowledgments, negotiate window sizes, and close connections. Each segment is divided into two parts, the TCP header and the data (See Figure 6-7).

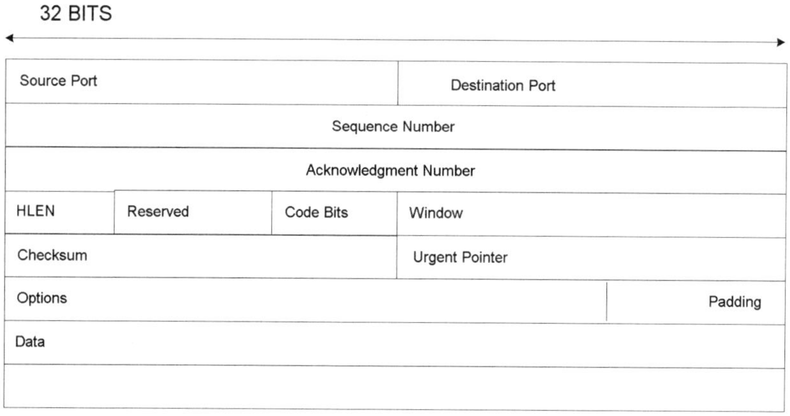

32 BITS

Source Port		Destination Port	
Sequence Number			
Acknowledgment Number			
HLEN	Reserved	Code Bits	Window
Checksum		Urgent Pointer	
Options			Padding
Data			

Figure 6-7. Transmission Control Protocol Format

- SOURCE/DESTINATION PORT: Contain TCP port numbers which identify applications at the ends of the connection.
- SEQUENCE NUMBER: Identifies the position of data.
- ACKNOWLEDGMENT NUMBER: Identifies the next expected octet of the stream from the other end.

- HLEN: Length of segment header.
- WINDOW: How much data the software is willing to accept.
- OPTIONS: Specifies various options such as maximum segment size.
- CODE BITS: Determines the contents of the segment, data, control, and connection establishment. These six bits tell how to interpret other fields in the header.

SUMMARY

TCP/IP is a layered architecture that does not correspond very well to the OSI Model. TCP/IP was developed long before OSI as a means of connecting minicomputers over a very wide area into what is called an internet, or, when speaking of the DARPA-sponsored network, the Internet.

Internet addressing was devised to provide a means of uniquely addressing any node anywhere on an Internet with tens of thousands of nodes. It is a 32-bit address, but is written in the dotted decimal form, for example, 144.49.7.3. To provide a system that is easier for users, nodes, computers, networks, and groups of networks on an Internet can be named hierarchically with a "domain" name such as johnd@company.com. TCP/IP provides "name servers" at various points in the Internet to translate domain names into Internet addresses.

Internet Protocol (IP), the lower of the two protocols, is responsible for the transmission of datagrams between nodes in the network. The service it provides is "best effort." It will not capriciously discard messages, but it makes no effort to correct errors. Included in the IP layer is the Address Resolution Protocol (ARP) function, which takes care of converting Internet addresses to the subnet addresses set for each node by the manufacturer of the network hardware. The Internet Control Message Protocol (ICMP) is an adjunct to IP. ICMP is used by IP when it needs to communicate control messages to a node which is not directly adjacent.

Transmission Control Protocol (TCP) is responsible for turning the connectionless, unreliable datagram service provided by IP into connection-oriented, error-free service, much like the Transport Layer of OSI. The User Datagram Protocol (UDP), on the other hand, is simply a way for "users," that is, application programs, to send and receive datagrams. IP provides datagram service only between nodes, not users.   The TCP/IP Applications/Application Services layer contains several widely used end-user applications as well as programs which provide services to application programs.

rlogin provides UNIX-to-UNIX remote command execution, allowing a user on one system to log in to another system on the network and execute UNIX commands on that node as if the user were on a locally attached terminal.

The Simple Mail Transfer Protocol (SMTP) provides for the transfer of e-mail messages between nodes on the Internet. SMTP is simple and is visible to users in the e-mail message headers. A number of "reader" programs are used to create and read messages.

TELNET is the virtual terminal protocol that most nongraphical TCP/IP applications use for terminal interaction. The term is also applied to terminal emulator programs that use the protocol. The TELNET emulator programs allow users of non-UNIX systems to log into a TCP/IP

UNIX node from a remote location and execute UNIX commands. The File Transfer Protocol (FTP) provides two connections between nodes. One is used for control commands, such as specifying the names of files on the remote node. The other is used for TCP stream data transfer in either direction between the nodes. The DOMAIN name server is used by applications to convert a hierarchical name to an Internet address. The DOMAIN name server will communicate with its peers in the Internet when necessary to convert a name for another domain. The Network File System (NFS) is a virtual filestore. It makes files which reside on remote nodes appear to the user and to application programs, to reside locally. NFS "exports" directories in a node, making them available to everyone on the network, or to selected nodes. Programs can "lock" a file or a record in a file so that two programs cannot simultaneously update it.

ADDITIONAL INFORMATION ON THE CD-ROM

- Additional information on terms and subjects found in this chapter
- Conversion Charts for Hexidecimal, Binary, and Octal Data Formats
- IP Traces
- ARP Traces
- UDP Traces
- IP RFCs
- TCP Traces
- ICMP Traces
- TCP RFCs
- TELNET RFC

7

SNA ARCHITECTURE

CHAPTER OBJECTIVES

This section takes a look at IBM's Systems Network Architecture (SNA). SNA was first introduced in 1974 and remains a central architecture in many of today's networks. The SNA architecture was designed to provide reliable, transaction oriented processing for large corporations. The objectives of this chapter are:

- Demonstrate the different types of Hardware used in SNA networks.

- Identify the layers of SNA and how they relate to the OSI Model.

- Discuss different SNA programs and their usage.

- Illustrate IBM's transition from a Master/Slave network to Peer-to-Peer Networking.

- Identify the major components of IBM's ESCON architecture.

IBM Hardware and Cluster Controllers

The IBM systems environment focuses on the mainframe CPU. The CPU "hosts" the application programs and, therefore, the communications software to support them. The CPU communicates with the outside world only via high-speed input/output (IO) channels which can traditionally be up to 400 feet long (the new ESCON fiber optic channels can reach nine kilometers), and the distance can be greatly extended with channel extenders. Controllers attach to the IO channels, as shown in Figure 7-1. Various peripheral devices, such as disk and tape drives are

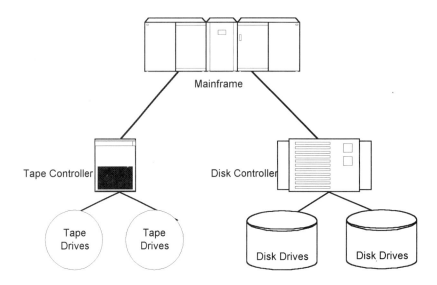

Figure 7-1. IBM Controllers

connected to the controllers. The principal purpose of a controller is to off-load the CPU. These relatively inexpensive systems take care of low-level peripheral control tasks that would otherwise use up valuable CPU cycles.

Figure 7-2 shows the simplest possible IBM data communications environment. A cluster controller is a device which controls a number of 3270-type displays and line printers. These devices attach via coaxial cable to the controller. Through a feature called the Asynchronous Emulation Adapter (AEA), ASCII devices, which emulate 3270, can also be attached. It is called a cluster controller simply because it controls a "cluster" of devices. The idea is to concentrate control logic for the devices into a single box, both to reduce costs and to allow one phone line and modem to serve several terminals.

The current IBM cluster controller, the 3174, was introduced in 1986. The 3274 is older but is still widely used. Throughout this chapter, when we refer to the 3174, we mean also to include the older 3274.

The 3174 cluster controller shown in this diagram is channel-attached. Although the terminals are shown directly attached, with the use of a multiplexer they can be located up to 3000 meters from the controller.

The 3174 is a micro coded device rather than a full-fledged computer and is capable of executing only the lower layers of the IBM data communications architecture, as we will explain later in this chapter.

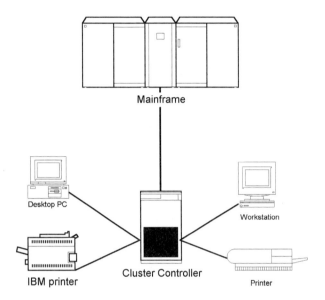

Mainframe

Desktop PC

Workstation

IBM printer

Cluster Controller

Printer

Figure 7-2. IBM Cluster Controllers

Communication Controllers and Concentrators

In order to communicate with the host CPU from a remote site, we must add another component. A 3174 can attach to phone lines through a modem. However, the CPU channel cannot be directly attached to a phone line, so a communications controller must be used. A communications controller interfaces and buffers data from remote devices such as cluster controllers or other communications controllers.

Figure 7-3 shows the simplest IBM remote data communications environment. When used as shown, a communications controller is referred to as a front end processor (FEP) because it contains a processor and operates at the "front end" of the processing of a data communications transaction (relative to the CPU at the "back end"). A phone line and modem connect the 3174 cluster controller to the communications controller, which connects to the CPU via the I/O channel. It is a programmable device, that is, it contains a computer. It runs highly specialized communications programs, not application programs. IBM communications controllers execute all but the highest levels of the SNA communications architecture.

The communications controller is either a 3720, 3725, or 3745. The 3720 is a lower

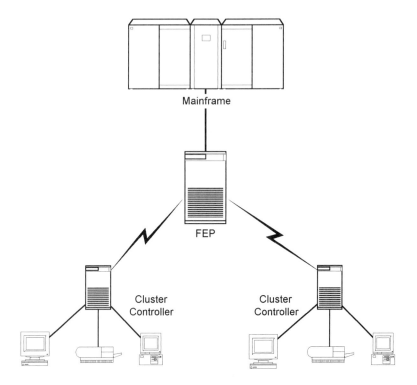

Figure 7-3. Front End Processors

capacity version of the 3725. The 3725 superseded the 3705, which was the workhorse of IBM communications for more than fifteen years. Over 48,000 3705s are installed.

The 37x5 can connect up to 786 remote cluster controllers, but each would require a separate communications line. It is often more economical to "concentrate" communications activity at a remote location onto fewer, faster links in order to reduce the cost of telecommunications lines. This is accomplished with a 37x5 at the remote site configured as a remote communications processor (RCP). The cluster controllers can be remotely attached to the 37x5, but can also be attached directly to the concentrator.

A 37x5 can communicate with more than one host, in which case it becomes an intelligent switch, meaning it can provide routing functions between cluster controllers and hosts. A 37x5 can perform all three of these functions—FEP, concentrator, and intelligent switch—at one time.

Through the remainder of this chapter, we will often use the terms FEP or RCP to mean any member of the 37x5 family of controllers when so configured.

The 3745 communications controller is the most powerful in the IBM line. The 3745 Model

210 contains a single central controller unit (CCU), providing about twice the power of the 3725. The model 410 contains two CCUs, which can operate in parallel to increase throughput or provide backup for one another in case of failure.

The 3745 family can communicate with the 3174 at speeds, ranging from 2400 baud to 64 Kbps. It can communicate with other communications controllers at speeds from 9.6 Kbps to 1.544 Mbps (T1 or E1). 37x5s also support devices attached through an X.25 packet switched network.

Throughout this chapter, we will use the term "37x5" as shorthand to mean the 3720, 3725, or 3745. For the purposes of this chapter, they are interchangeable.

Other 37x5 Functions

Two FEPs can communicate directly, interconnecting two hosts as shown. This configuration would be used to interconnect two SNA networks remotely, such as networks in two cities (See Figure 7-4).

An FEP can attach directly to the channels of more than one host. It will route messages it receives on its communications lines to the appropriate host. This is one way in which the CPUs can communicate directly.

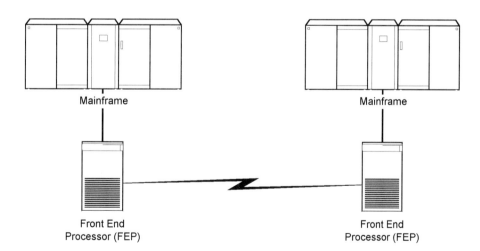

Mainframe   Mainframe

Front End   Front End
Processor (FEP)   Processor (FEP)

Figure 7-4. FEP to FEP Connectivity

IBM Token Ring Features

In 1987, IBM began providing Token Ring access to the host for local users via one of two gateways (See Figure 7-5):

1. Channel-attached 3174 cluster controller with the Token Ring 3270 Gateway feature. With this feature installed, a 3174 can act as a gateway to an SNA network for other 3174 cluster controllers in its ring. The gateway must also have the Token Ring Connectivity features, which enable the 3174 to connect to a Token Ring LAN. Devices attached to the Token Ring, called downstream nodes (DSN) and including PCs, 3174s, and AS/400s, are provided host access via the gateway. SNA is used to provide this communications capability from DSNs to the host.

2. Channel-attached 37x5 with Token Ring Interface Controller (TIC). The TIC allows the 37x5 to attach directly to a Token Ring. The 3174s on the ring can then use the 37x5 to communicate with the host CPU or other parts of the network.

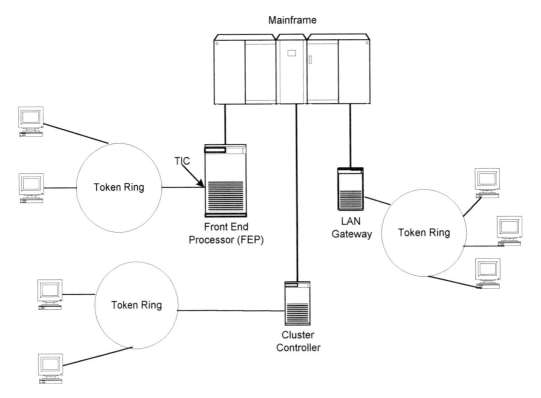

Figure 7-5. Token Ring Host Connectivity

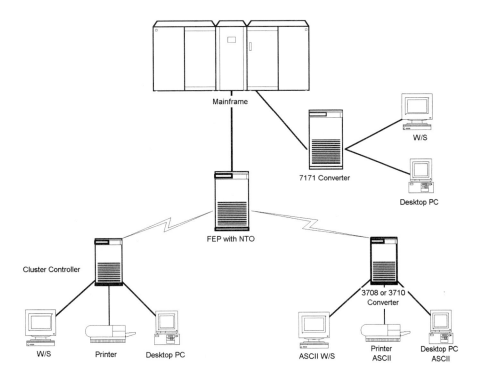

Figure 7-6. IBM Protocol Converters and Concentrators

Data flow from terminals through downstream physical units to upstream physical units and then to the host through the FEP.

As mentioned above, ASCII devices, including PCs emulating ASCII devices, can attach to 3174s that have the Asynchronous Emulation Adapter (AEA) feature. But if only ASCII devices or other non-3270 type devices need to attach to an SNA network, the IBM 3708 Network Conversion Unit can provide that capability at a lower cost than the 3174. It is a protocol converter which can connect up to 8 ASCII devices to a 37x5, making them look to SNA as if they were on a 3174 cluster controller. The 3708 can also connect to a non-IBM host, allowing the devices alternatively to communicate with the non-IBM host in their native mode (See Figure 7-6).

The IBM 3710 Network Controller provides the capabilities of the 3708 but supports up to 56 devices and attaches to as many as five hosts.

The IBM 7171 Protocol Converter attaches directly to the IBM channel, rather than to a 37x5, supporting up to 64 ASCII devices and looking like a channel-attached 3274 controller.

IBM also incorporates protocol conversion in the 37x5 with the Network Terminal Option (NTO). NTO software converts ASCII data streams to SNA-compatible data streams.

SNA Overview

Having explained the hardware environment of IBM data communications, we will now turn to the software environment—Systems Network Architecture (SNA).

Different requirements have produced quite different communications architectures. As you have seen, TCP/IP was designed to interconnect many researchers and similar users for the exchange of information over a wide area. DECnet was designed for resource sharing by networked minicomputers. SNA was designed for transaction processing in mission-critical applications, often involving services provided to customers. SNA networks usually involve a large number of terminals communicating with a mainframe. Typical transactions perform inquiries and update information in a database. For example, a commercial bank might have a number of 32703270-type display units and printers in each of hundreds of branch offices which are used to access a central database in the home office. The requirements for transaction processing are:

- Responsiveness. Transaction requests and responses must be transferred between terminals and the host CPU without delay.

- Predictability. Responsiveness must not suffer appreciably under conditions of maximum load.

- Reliability. Reliable, error-free transmission of requests/responses must be provided.

These requirements, together with the underlying hardware environment, explain two important features of traditional SNA:

1. Traditional SNA Is Connection-Oriented

In a connectionless environment such as TCP/IP, the path a message might take is unpredictable. There is also overhead for each message at each intermediate point to determine its routing. On the other hand, in a connection oriented environment, the path is determined when the connection is established and will not change. Alternate paths can be defined to allow for path failures. Thus, each intermediate node can route messages with minimal overhead once the connection is established.

Furthermore, because of the nature of the hardware in an IBM environment, the establishment of connections requires the involvement of the host communications software. Since traditional SNA is connection-oriented, the host need not be involved once the connection is established. If it were connectionless, the host would need to be involved every time a message was transmitted.

2. Traditional SNA Is Hierarchical

In an SNA network, data communications functions are performed by a hierarchy of hardware with varying levels of programmability, from full programmability of the CPU to minimal programmability in the controllers. As you have seen, not all hardware elements are capable of executing all layers of the SNA architecture. This environment dictates a hierarchical "master/slave " control scheme. The CPUs are the only components capable of executing the higher layers of the protocol, that is, the only nodes capable of assuming the role of "master." One important example is the establishment of connections between nodes. Therefore, the CPUs alone are capable of exercising control of the network. In fact, every traditional SNA network must contain at least one host CPU.

In other chapters we have treated "nodes" simply as devices which contain a computer and attach to the network via communications ports. In SNA terminology, we must modify this definition somewhat. Host CPUs, communications controllers, and cluster controllers are all nodes, so a node might or might not contain a computer. And, as you have seen, the host CPU communicates with controller nodes via its channels—it does not have a communications port of its own.

Once again, it will be helpful to refer to the service/protocol model to understand the fundamental building blocks of SNA. A node provides certain services to the user of the node, typically oriented toward the input or display of data. It uses a protocol to communicate with the network in order to provide its services, and that protocol is specific to the node, designed to suit the service that the node provides.

SNA Node Services and Protocols

The service/protocols of a node are determined by several things: the type of the node, the logical units (LUs) it contains, and whether or not it is designated as a System Services Control Point (SSCP).

SNA nodes are divided into several types:

- Peripheral Nodes are controllers.
- Type 1 nodes are the older cluster controllers such as the 3276, the predecessor of the 3274.
- Type 2 nodes are newer cluster controllers such as the 3274 and 3174.
- Subarea Nodes are the 37x5s and the CPUs.
- Type 4 nodes are the 37x5s.
- Type 5 nodes are the CPUs.

Although the original definition of SNA provided for a Type 3 node, that designation has never been used.

Data Stream Protocols .

SNA defines four types of data stream protocols:

1. SNA Character String Data Stream. This is EBCDIC data interspersed with control codes which can be used to format the character string, for example, for printing on a printer. This is often referred to as the 3767 data stream protocol for use with RJE devices.

2. SNA 3270 Data Stream. This contains characters interspersed with control codes for the 3270 Display Unit.

3. SNA 5250 Data Stream. This contains characters interspersed with control codes for the 5250 Display Unit.

4. User-Defined Data Streams. Typically used for communication between programs in Type 5 nodes.

The data stream protocol used between two nodes is determined by the nature of the most powerful node.

PUs and LUs

Each SNA node is said to contain a physical unit of a certain type and one or more logical units of a certain type (See Figure 7-7):

- Physical Unit (PU). Each node in an SNA network contains at least one PU. The physical unit implements physical connections to the network. SNA defines PU types that correspond directly to the node types above: PU 1, PU 2, PU 4, and PU 5. Thus, a Type 2 node, for example, is said to contain a PU 2.

- Logical Unit (LU). Each node contains one or more LUs. For a cluster controller, each device in the cluster (each terminal, printer, etc.) has one LU. The function of the LU is:

The LU has several functions. In the cluster controller, it implements data transfer between the PU (the controller) and the device (terminal, printer, etc.). In the host, it implements data transfer between the PU and the application. In the controller, it implements data transfer between communication lines, or between a communications line and the CPU channel.

The type of the LU defines the type of data that the LU is capable of accepting and the services it provides. These are the LU types originally defined for SNA:

LU 0. This is the "catch-all" type in nodes that do not fit any of the following descriptions. Uses User-Defined Data Stream.

LU 1. Used in a device such as a printer. Uses SNA Character String Data Stream.

LU 2. Used in an IBM 3270 or any other similar cluster controller. Uses the SNA 3270 Data Stream.

LU 3. Used in a printer that is not attached to a cluster but which uses the SNA 3270 Data Stream.

LU 4. Like LU 1, but supports direct communications between LUs in the same cluster controller.

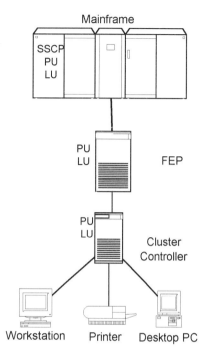

Figure 7-7. SNA Datastream.

These five LU types were all that were defined for the original version of SNA. As you can see, there is no explicit LU type for program-to program communications. The final two LU types were added later to accommodate this need and the advent of the 5250.

LU 6. Used in CPU nodes for application programs to allow program-to-program communication. Uses any of the defined data streams, but usually the User-Defined Data Stream.

LU 7. Used in a 5250-type display with the corresponding data stream.

Like the PU 1, LU 5 has never been defined. Two successive versions of LU 6 have been defined: LU 6.1 and LU 6.2. You will learn about these when you study program-to-program communications later in this chapter.

System Services Control Point (SSCP)

If a node contains an SSCP, it provides the services needed to manage an SNA network. Functions of the SSCP include:

- Establishing connections between LUs
- Allocating resources needed to establish connections
- Providing control of the network to the network operator (the user who controls operation of the network)
- Converting symbolic network names to actual addresses
- Managing error recovery within the network

The Logical Hierarchy of SNA

It is important to understand how the building blocks you have seen are arranged into the SNA hierarchy (See Figure 7-8).

Beginning at the top, the hierarchy consists of:

Network. All interconnected nodes taken collectively.

Domain. Defined by a PU 5 node (a CPU) which contains an SSCP. Includes all nodes managed by that SSCP.

We said earlier that an SNA network must contain at least one CPU; in fact, it must contain at least one SSCP, which can exist only on a PU 5 node. A network therefore can contain as many domains as it has PU 5 nodes.

Subarea. Defined by a PU 4 or PU 5 node (37x53745 or CPU). The number of subareas equals the number of PU 4 and PU 5 nodes. The subarea includes all PU 2s to which services are provided.

End Nodes. Each PU 1 or PU 2 node (cluster controller) in a subarea.

The SSCP in each domain knows about the nodes in that domain, but not about the nodes in other domains in the network. The Cross Domain Resource Manager (CDRM), which is part of the SSCP, knows the characteristics of nodes in other domains, even though they are managed by a different SSCP, and it takes care of establishing connections that cross domains.

As you can see, SNA network nodes are positioned in the hierarchy according to their capabilities, with nodes containing computers always at the top, acting as "masters," and nodes with little or no intelligence at the bottom, acting as "slaves."

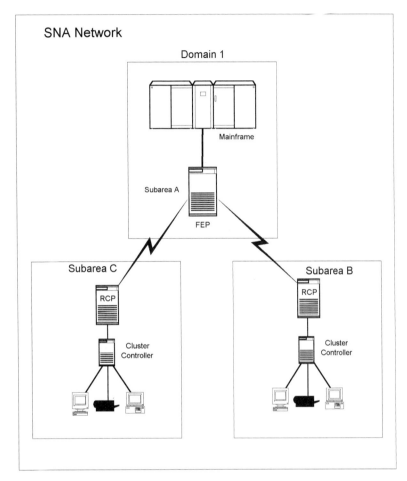

Figure 7-8. SNA Network

Network Addressable Units

The communication of data in a SNA network can be considered to fall broadly into two parts (See Figure 7-9) First, the connection is established between the nodes participating in the transmission and managing the transmission from end-to-end. Second, data is transmitted between points in the network on the path established for the connection.

NAU and PCN Architecture

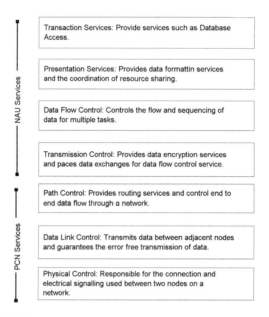

Figure 7-9. NAUs and PCNs

The components of SNA are divided along those same lines:

Network Addressable Units (NAUs). The units you have seen, PUs, LUs, and SSCPs, are collectively called Network Addressable Units because they have addresses and can communicate with one another. NAU Services are simply the services provided by the NAUs in the nodes, together with the Transmission Control, Data Flow Control, and Function Management layers of SNA. NAU services include:

•   Establishing connections
•   Managing the end-to-end flow of data across the connections
•   Providing certain application and user services
•   Overall management of the network

The Path Control Network (PCN). The Path Control Network consists of the components that move data across the network from point to point. PCN Services are implemented by certain hardware components and microcode within the nodes, and by the Path Control and Data Link Control Layers of the SNA architecture. PCN Services include transmission of data across a specific physical circuit, including routing of data from node to node on an established connection and segmenting and blocking of messages.

Every SNA node must contain a standard set of PCN functions. Although the implementation of the functions can vary widely depending on the hardware environment, the functions are common to every SNA node.

In this section we will look at the broad functions of NAUs and the PCN. In the next section we will look more closely at the corresponding function layers of the SNA architecture.

SNA Sessions

In SNA, connections take the form of sessions. Sessions take place between SNA nodes. But, as you have seen, a node such as a cluster controller can have many devices, such as 3270-type display units and printers, attached to it. Because more than one device will want to use the network at any one time, there must be a way for several sessions to take place at the same time (See Figure 7-10).

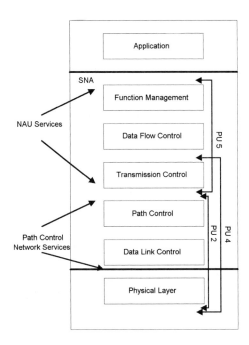

Figure 7-10. NAUs and Sessions

LUs resolve this issue. Application sessions take place between LUs. Each application subsystem can have many LUs in the host CPU.

Since nodes contain multiple LUs, multiple sessions can be established over a single communication line between nodes.

Sessions also are established between two SSCPs in order to set up cross-domain communications, and between the SSCP and PUs and LUs in the domain to control the network and to establish LU-to-LU sessions.

Network Addressable Unit (NAU) Services

The functional components contained in the various NAUs provide everything apart from the actual transfer of data between nodes required to establish and conduct SNA sessions. These higher level functions include:

- Establishing and ending sessions
- Managing the network
- Network operator service
- Presentation services such as formatting of data
- Application services such as synchronization of activities

The establishment of sessions between LUs and the assignment of routing priorities are primary functions of the SSCP, which, you will recall, is one type of NAU. The SSCP resides in a PU 5 because the intelligence of the PU 5 node is required to establish and maintain the session. Once the session is established, communication between the LUs requires only the LUs themselves, and data need not flow through the CPU.

There are specific protocols and commands that are invoked when a host application program (or an end user) desires to communicate. The physical unit (PU) in the cluster controller must first be activated. This is followed by activating one or more logical units (LUs) which essentially represent a device connected to the controller (such as a terminal or PC running an emulation program). A series of SNA commands follow which connect the end user to the application program and, most importantly, set the rules to be adhered to while in session. These rules are set up as session profiles when the PU, and especially the LUs, are defined in VTAM. The rules are sent to the LU in a BIND command, which is the SNA command that actually establishes the session.

For example, a display station LU is configured to follow the rules for a LU Type 2. The rules are broken down into three major categories:

1. Transmission services (TS) profiles—used to specify pacing algorithms (flow control), sequence number usage (tracking of sent and received frames to ensure no frames are lost), and maximum message lengths that are allowed between the host and the LU.

2. Function management (FM) profiles—used to specify whether the LU must respond immediately to a request made by the host, the SNA commands that will be sent (such as SIG, CANCEL, LUSTAT, BID, etc), chaining usage (a unit of data recovery), bracket usage (an uninterruptable unit of work such as needed to acquire a printer for exclusive use), and recovery responsibilities.

3. Presentation services (PS) profiles–used to specify the application-level commands supported, such as queries to the device and screen sizes allowed (1920, twenty-four rows, eighty characters/row).

Path Control Network (PCN) Services

The PCN handles the transmission of messages in the SNA network. Its responsibilities include:

* Routing messages according to the connection established by the NAUs
* Transmission of data, possibly over parallel links established for the connection, including segmenting and blocking, sequencing, pacing, and error correction

SNA Functional Layers

NAU services and PCN services encompass the SNA functional layers. The SNA functional layers relate more directly to the architecture of SNA in the sense that the term was used previously in this course.

SNA Layer Services

The services provided by each layer will seem familiar by comparison with the OSI Model. Recall that the physical and application layers are considered to lie outside SNA. Services of the SNA layers are:

* Data Link Control Services. Transmission of data across a specific link, including error detection and retransmission of frames when necessary.

* Path Control Services. Routing of messages according to route established for the session, parallel transmission of messages over multiple links, addressing of messages between subareas.

- Transmission Control Services. Pacing of messages to prevent overrunning a slower node, checking sequence of messages to detect lost messages, encryption and decryption.

- Data Flow Control Services. Maintaining the integrity of data flow within a session for entire messages.

- Function Management Services, consisting of end-user services, which are concerned with the exchange of data between LUs and include the formatting and display of data and synchronization of activities.

To see how the SNA Headers correlate to the OSI Model headers, see Figure 7-11.

Session network services, which are the services of the SSCP, include:

- Activation and deactivation of sessions
- Configuration of the network
- Handling of errors and facilities in the network
- Providing communication between the SSCPs and the network operator

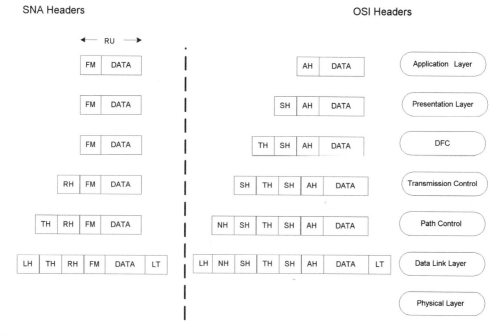

Figure 7-11. SNA Header Format

SNA Message Format

SNA is very much like OSI with respect to the formatting of messages as they flow between layers. Data units in SNA are:

- Message. The highest level, the actual user data.
- Request/Response Unit (RU). A message or possibly a portion of a long message that has been divided into parts and has been encapsulated by the Function Management Layer. RUs implement the LU-to-LU protocol.
- Basic Information Unit (BIU). An RU encapsulated by the Transmission Control Layer.
- Path Information Unit (PIU). A BIU encapsulated by the Path Control Layer.
- Basic Link Unit (BLU) or SDLC Frame. A BIU encapsulated for transmission via SDLC (an HDLC subset) by the Data Link Control Layer.

SNA Programs

Many of the services provided by the SNA functional layers are provided by software on the nodes which are programmable, that is, on Type 4 and 5 nodes (and Type 2.1 nodes, which will be discussed later). Unlike TCP/IP nodes, where the software is essentially part of and included with the operating system, SNA software exists in separate "program products" for which there is a separate charge in some cases. This section will introduce some of the more important IBM program products (See Figure 7-12).

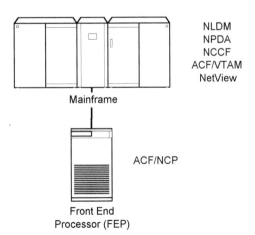

Figure 7-12. SNA Programs

As you have seen, different nodes have different programming capabilities, so there are necessarily different program products for them. When IBM introduced SNA, program products which previously supported older protocols began to support SNA and were given the prefix ACF (Advanced Communication Function), to signal their support for SNA.

For Type 5 nodes (CPUs), the important programs are VTAM, NPDA, NCCF, NLDM, and Netview, described below.

ACF/VTAM (Virtual Telecommunications Access Method). VTAM implements the PU, LUs, SSCP, and PCN on the CPU. It is called virtual because of its capability to establish multiple virtual circuits on a single communications link. See Figure 7-13 for a simple SNA VTAM configuration.

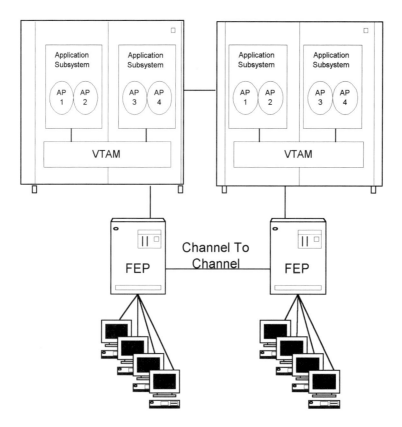

Figure 7-13. VTAM

The Major elements of VTAM are:

- VTAM— the actual host program which implements the SNA Architecture
- VTAM Application Programs, such as CICS and IMS
- NCPs
- Physical Units and Logical Units
- Data Links

Four of the major functions of the VTAM system include:

1. Starting and stopping the network
2. Allocating resources in the network
3. Dynamically changing network configurations
4. Managing exchange of data between VTAM application programs and logical units

VTAM Application Programs. Any program that uses VTAM macro instructions is considered a VTAM application program. CICS (Customer Information Control System), IMS (Information Management System), and any user-written VTAM application are some examples. These programs are written in assembler and make "calls" to VTAM with macros such as OPEN, CLOSE, OPNDST, CLSDST, SEND, and RECEIVE.

The responsibilities of VTAM application programs are:

1. Connecting/disconnecting to/from VTAM
2. Initiating and terminating sessions with logical units
3. Sending and receiving to and from logical units
4. Controlling data flow

NCCF (Network Communication Control Facility). NCCF provides control to the network operator of multiple-domain networks, allowing messages to be sent between operators in different domains, providing assistance with problem determination, etc.

There are two programs which run under the control of NCCF:

NPDA (Network Problem Determination Application). Provides more sophisticated problem determination tools.

NLDM (Network Logical Data Manager). Provides data on active sessions to assist the network operator with problem determination.

NetView. Replacing NCCF, NetView provides the capabilities of the NCCF and its adjuncts while presenting a better interface to the network operator and incorporating some features oriented towards the newer elements of the IBM communications environment such as Token Ring. NetView serves as the domain manager, whose functions include collecting statistics on any node under its control that is capable of sending certain types of SNA messages called Management Services RUs. These messages are sent on the PU to PU session and are routed to the NetView application by VTAM. An ALERT is a term used to describe an event, such as an error condition occurring on a Token Ring, that is forwarded to NetView and logged in a database. The operator can then call up (or counters and thresholds can be set to automatically display) data on the state of the various network elements.

For the Type 4 nodes (37x5s), important programs include:

ACF/NCP (Network Control Program). NCP is a table-driven control program which implements SNA protocols interacting with an access method (such as ACF/VTAM) to control a network. It offloads the mainframe from having to consume CPU cycles in managing and controlling much slower devices such as modems.

Some of the control functions include:

- Dialing as well as answering stations in a switched environment
- Polling and addressing stations in dedicated (leased) environments
- Maintaining error data for the attached devices and the communications links
- Managing buffer storage dynamically as data is sent and received
- Selecting data link line speeds

Although the 37x5 is not programmable in the usual sense, the systems programmer writes a set of table statements which describe the network. These statements are actually System 370 Assembler Language macro statements. At the CPU site, the programmer "generates" the NCP code by assembling the macro statement on the CPU. The result is binary code which is the NCP (often called an "NCP load module"). It is then "downloaded" from the CPU to the 37x5.

NCPs require a great deal of maintenance. A macro statement is required for each line, PU and LU, for which the 37x5 is responsible, and these macros must be modified and a new NCP generated and downloaded every time a change is made to the subarea or the links to other subareas change. However, SNA does offer certain host features that allow a more dynamic configuration of network node resources such as LUs. This functionality is implemented in VTAM and is called Dynamic Definition of Dependent LUs (DDDLU). This particular feature requires that the PU be set up in such a way (through PU definition statements in VTAM) that it will activate an LU on the host whenever a device is connected and powered-on by the user.

SNA Peer to Peer

Since SNA's introduction in the mid—1970s, the microprocessor has been invented. At one time, a large network might have only a single node containing a general purpose computer (the host mainframe). Now, in most networks, virtually every node is either a personal computer, workstation, or network server, and so contains a microprocessor. Although this change in the hardware environment has many implications for the architecture of networks, it creates one difference that required extensive enhancement of SNA: it became necessary for programs within these peripheral nodes to communicate. For example, the users of workstations wanted to share data using NFS.

IBM did not respond to this changing environment with a rearchitecting of SNA, which would have required a wrenching conversion for the tens of thousands of SNA networks already in existence. Instead, it added enhancements to SNA which left the previous capabilities of SNA intact. This also maintained the layered architecture, yet provided the enhancements necessary for the new "intelligent" peripheral nodes such as IBM's RS 6000 workstations on a Token Ring LAN.

APPC

IBM accomplished this with the introduction of the Advanced Program-to-Program Communication (APPC) facility. APPC consists, essentially, of two new NAUs:

LU 6.2. A new LU type which supports program-to-program communication, including communication between programs in peripheral nodes, for example, file transfers between workstation database programs. It supports both hierarchical and peer-to-peer communications. It is the only LU type supported by APPN.

PU 2.1. A new PU type which supports peer-to-peer communication between peripheral nodes, such as a PC emulating a 31743274 cluster controller.

As you have seen, the original design of SNA assumed that sessions would take place between programs in CPUs and terminals attached to cluster controllers. A session would typically be between LU 2s in both nodes. This constituted a "master/slave" relationship, with the CPU node having responsibility for functions such as error correction.

LU 6.2 enables peer-to-peer communication between PU 2.1 peripheral nodes, or between a PU 2.1 peripheral node and a host CPU. Note that the communication is peer-to-peer only from the point of view of the SNA network. The relationship between the cooperating programs in a peer-to-peer session can take some other form, such as client/server or even master/slave. That is determined by the design of the programs.

Figure 7-14 shows how application programs can interface with underlying transport layer protocols. Application programs use the Common Programming Interface for Communications (CPIC) interface in order to connect to different types of network topologies and protocols. Applications using the CPIC interface can send data over TCP/IP, SNA, NetBIOS, or Open Systems architectures.

LU 6.2 provides these additional capabilities to the communicating programs:

- Allocation of Primary Node Designation. Either node can be the "primary node" to take responsibility for error handling.

- Synchronization. Provides the nodes with the facilities necessary to synchronize their activities.

- Parallel Sessions. Allows multiple sessions to be established for a single pair of LUs. With parallel sessions, the overhead of establishing and managing separate LUs for each session is avoided.

APPC allows an SNA network to look much more like the OSI and TCP/IP networks you have previously studied, where the relationships throughout are peer-to-peer, rather than the master/slave relationships that previously prevailed. Note, however, that a PU 5 node is still required for each network in order to run the SSCP, and the SSCP must establish the peer-to-peer session.

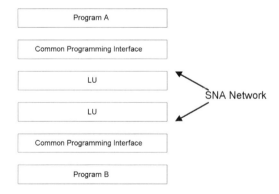

Figure 7-14. CPIC Interface

APPN

APPN stands for Advanced Peer-to-Peer Networking. It was developed by IBM and provides networking services similar to that of the TCP/IP protocol stack. It is a network layer service and can run under a variety of data link and physical layer protocols such as Ethernet, Token Ring and FDDI. There are additional features of APPN networks:

- APPN was first available in 1986 on the System/36, 1988 on the AS/400, and 1991 on the 3174 and OS/2.
- It is made up of interconnected Type 2.1 nodes.
- The backbone is made up of interconnected APPN network nodes.
- End nodes attach to the backbone network node servers.
- APPN provides routing for APPC sessions (Peer-to-Peer) through 2.1 nodes.
- Dynamic update of network node routing tables.
- Directory Services are provided.
- It uses the SNA LU 6.2 protocol.

APPN consists of the three primary components, The Low Entry Node, the End Node, and the Network Node. Each of these nodes provides a different level of functionality. These components are shown in Figure 7-15 and are described below.

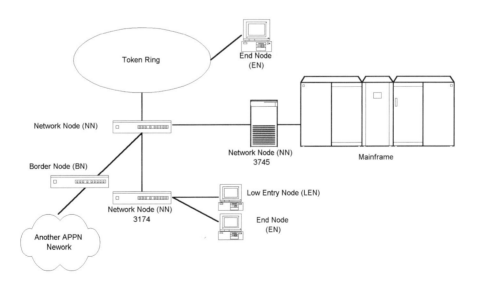

Figure 7-15. APPN Network

Network Nodes are nodes which route data traffic between End Nodes. Network Nodes send routing and topological information to each other in order to stay current when changes are made in the network. An example of a Network Node is a 3174 Cluster Controller.

End Nodes are computers on the network which have their own operating system and can send and transmit data to the network. When End Nodes are turned on, they transmit data to their attached Network Node which stores information about each End Node attached to it. The Network Node takes care of session establishment and traffic routing for the End Node.

Low Entry Nodes require Network Nodes to communicate in the network. An example of this type of device would be a PC running DOS. PCs running OS/2 have End Node capabilities built in to the operating system.

Border Nodes are nodes which route traffic between individual APPN networks. Border nodes provide isolation of local traffic and routing of internetwork traffic.

Within the Peer-to-Peer framework of APPN a client/server relationship also exists. The client/server relationship exists between End Nodes and Network Nodes, and between Low Entry Nodes and Network Nodes. Services for the three different nodes types are:

- LEN: Adjacent node communications only. Resource definitions must be manually configured.
- End Node: Provides directory and routing services. Supports LU 6.2 sessions with Network Node Servers.
- Network Node: Provides server functionality such as directory services and routing.
- Provides support for local LU sessions and intermediate session routing.

ESCON Components

An ESCON communication system consists of a number of different components, each listed and described below. Refer to the ESCON components diagram (Figure 7-16) for the following discussion.

The ESCON Channel is the host-based channel connection that supports the ESCON architecture. It directs the flow of information between IO devices and host main storage and provides common controls for the attachment of different IO devices. IO devices are attached through control units to the channel subsystem via channel paths.

An ESCON control unit is any control unit (cluster controller, tape drive, DASD, printer, or other similar devices) that can connect directly to an ESCON Channel or Director port. A non-ESCON (or bus-and-tag) control unit cannot connect directly to an ESCON Channel or Director, but must interface through an ESCON Converter. A control unit provides the logical capability necessary to operate one or more IO devices and can be connected to more than one channel at the same time over different link interfaces.

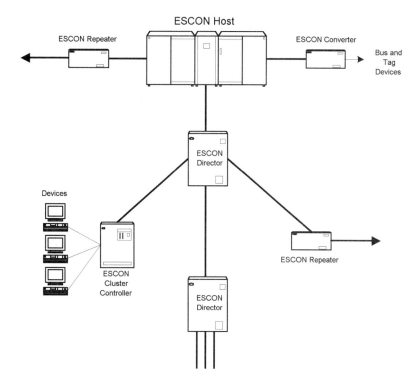

Figure 7-16. ESCON Components

A control unit image has the logical appearance of a control unit. Each control unit image appears to be an independent control unit, although these images share the same link-level facilities and physical paths. Up to 256 IO devices can be attached to each control unit or control unit image.

The ESCON Converter is intended as a migration support tool and is available in two models. The 9034 Model 1 allows parallel bus-and-tag devices to communicate with the serial ESCON channel and can be attached directly to the host or to an ESCON Director. The 9034 Model 2 allows ESCON-compatible DASD devices to communicate with the bus-and-tag channel.

The ESCON Repeater regenerates bi-directional ESCON signals on full-duplex ESCON links to extend the distance between an ESCON channel and control unit or another ESCON channel.

Channel to channel connections as long as 60 km are possible with two chained repeaters on a single-mode link. Connections as long as 9 km are possible using two chained repeaters on multimode links. The maximum extension distance depends on devices, such as directors and types of controllers, installed in the system and fiber cable quality.

The Director, sometimes referred to as a "dynamic switch," is a switch that acts as a communications hub for ESCON channels. It provides the capability to physically interconnect any two links that are attached to it. Such a connection between two ports provides simultaneous, two-way information transfer.

When a connection is established, the two ports and their respective point-to-point links are connected so that frames received by one of the ports are passed transparently to the other port. Such a connection can be either static or dynamic. A static connection between two ports is set up and removed by the Director itself. A dynamic connection is set up and removed by delimiters or sequences in the flow of data through the ports; that is, when a connection is to be made, addressing information in the frame determines to which port the incoming data should be routed. Start-of-frame (SOF) and end-of-frame (EOF) delimiters are used to initiate and remove dynamic connections.

An internal Director port is used for communicating configuration and control information with the channel. This port has the appearance of a control unit, is assigned a link address, and is referred to as a dynamic switch control unit. The dynamic switch control unit allows creation and removal of static connections, placing of ports off-line, and modifications to dynamic connections.

Multiple channels can be attached to a Director to allow sharing of IO devices attached to the same Director.

The Director is currently available in three models, the Model 1 (supporting up to sixteen links), the Model 2 (supporting up to sixty links), and the Model 3 (supporting up to 124 links). The ESCON System Connection Manager, or ESCM is host-based system software that supports and allows configuration of ESCON channels.

An ESCON fiber link is a full-duplex, point-to-point, fiber-optic connection between a control unit and a channel, a channel and a Director, a control unit and a Director, or a Director and another Director. A link is attached to a channel or control unit through the link interface of that channel or control unit, and to a Director through one of its ports.

Physically, there are two types of fiber links:

1. Jumper cable—This connects an ESCON Channel or ESCON device to a distribution panel, or to each other. Jumper cables can use either 9-10/125-micron or 62.5/125-micron optical fiber.

2. Trunk cable—This contains multiple optical fibers and is used to connect one distribution panel to another. It can use either 9-10/125-micron, 62.5/125-micron, or 50.0/125-micron optical fiber.

SUMMARY

To understand SNA, one must be aware of the hardware environment for which it was designed. IBM data communications focuses on the mainframe CPU, which communicates only over its IO Channel. For data communications at the CPU site, the 31743274 cluster controller attaches to the CPU channels and supports many 3270-type display units, printers, or PCs or workstations emulating 3270s. Remote data communications requires controllers such as the 37x5, which attaches directly to the CPU's IO channels to support possibly hundreds of communications lines. Remote 3174s communicate through the 37x5 acting as a "front end processor" (FEP) to the host CPU. The 37x5 can also act as an intelligent switch and a concentrator in remote sites.

SNA was designed for transaction processing in mission-critical applications. Principal benefits of SNA are responsiveness in handling transactions, predictable transaction throughput rates even under load, and reliable data communications.

SNA is hierarchical and connection-oriented. Its hierarchical arrangement of nodes makes the intelligence of the CPU available for establishing connections and managing the network, but allows data transfer to take place without the use of valuable CPU cycles. SNA's connection-orientation is well-suited to transaction processing. The overhead for routing of messages is minimized because they flow along paths defined by the connection. Connection-oriented service is much more predictable and stable than datagram service.

The IBM SNA environment is different from other communications environments in that not all nodes in the network can run all layers of the communications architecture. Only the host CPU can run all of the highest layers of SNA, so it must participate in establishing connections and in all network management functions.

SNA defines physical units (PU12 s) in each node which implement the transfer of data to and from the network. Logical units (LUs) implement data transfer from the PUs to the actual source or destination of the data within the node. System Services Control Point (SSCP), which can exist only in a CPU, manage the establishment of connections between LUs. Since every SNA network must have an SSCP to manage it, every network must have a CPU.

Data transfer in a SNA network falls broadly into two parts: the establishment of the connection and the transfer of data across those connections. Establishing connections is the responsibility of the network addressable units (NAUs), which include the PUs, LUs, and SSCPs. Data transfer is the responsibility of the Path Control Network (PCN).

The functional layers of the SNA architecture implement the NAUs and the PCN. The data link control and path control layers provide PCN services, and the transmission control, data flow control, and function management layers provide NAU services. SNA was the predecessor to OSI and is very similar with respect to its layering. Like any layered communication system, SNA encapsulates data with headers, surrounding the actual data bytes with necessary control information.

Two IBM program products are required to run SNA on IBM hardware: The Virtual Telecommunications Access Method (VTAM) on the host CPU, and the Network Control

Program (NCP) on the 37x5. NetView runs on the CPU above VTAM to provide network management services.

The advent of the microprocessor led to the enhancement of SNA to provide for communication between application programs running on intelligent non-host CPU nodes, called peer-to-peer communication. The Advanced Program-to-Program Communication (APPC) facility created new logical and physical unit types which allow the participating nodes to take over some of the responsibilities that previously belonged to the host CPU, such as error correction, and to provide services needed by communicating programs, such as synchronization.

ADDITIONAL INFORMATION ON THE CD-ROM

- Additional information on the ESCON protocol
- SNA Traces
- SDLC Traces
- ESCON Traces

8

DNA ARCHITECTURE

INTRODUCTION TO DNA

Because DNA is, with its latest version, OSI compliant, very little will be said about the communications architecture itself in this chapter. We will, however, introduce you to some unique DEC hardware configurations and specialized communications protocols which are important in order to understand the DEC environment and the suite of DECnet applications.

The term Digital Network Architecture (DNA) refers to the definition of the communications architecture. DNA has been implemented by DEC and by others. DEC's implementation is called DECnet. The term DNA will be used in this section, unless speaking specifically of the DECnet implementation.

CHAPTER OBJECTIVES

At the completion of this chapter, you should be able to:

- Differentiate between tightly coupled multiprocessing (MP) and loosely coupled multiprocessing as found in Local Area VAX clusters (LAVCs) and VAXclusters

- Indicate the important features and components of Local Area VAXclusters and VAXclusters

- Identify the features of System Communication Architecture (SCA), Local Area Transport (LAT), and Digital Network Architecture (DNA)

- Name the protocol used to download operating software

- State the primary differences between DNA Version IV and DNA Version V

- Identify the proprietary DEC protocols used by the Data Link layer of DNA

- Identify the proprietary DEC protocols and services related to the Network Application Layer of DNA

DEC PRODUCT OVERVIEW

DEC computers employ two computer architectures (See Figure 8-1):

- Virtual Address extension (VAX). This is DEC's proprietary complex instruction set computing (CISC) architecture.

- MIPS architecture. This is a reduced instruction set computing (RISC) architecture developed by MIPS Computer Systems, Inc. DEC usually refers to this simply as their RISC architecture. DEC's implementation of the MIPS architecture reverses the byte ordering defined by MIPS.

The VAX products are further divided into two groups:

- MicroVAXes, which are based on microprocessors

- VAXes, which are based on board-level computers

Technology	Desktop Products	Multi-User Products
VAX	N/A	VAXes
Micro-VAX	VAXstations	Micro-VAXes
RISC (MIPS)	DECstations	DECstation server configurations

Figure 8-1. DEC Product Family

All VAX products run a version of Virtual Memory System (VMS), DEC's proprietary operating system, and can also run DEC's version of 4.3 BSD UNIX, called ULTRIX. ULTRIX complies with the U.S. Federal Government POSIX standard. The MIPS-based products run only ULTRIX.

DEC sells two lines of workstation products. Both lines can be configured either as single-user desktop units or as multi-user network servers:

- VAXstations, which are MicroVAX/VMS systems

- DECstations, which are MIPS / ULTRIX systems

VAX minicomputers include several families of processors offering a broad range of performance and capacity. Some VAXs have multiprocessor, or multiple-CPU architectures. This type of multiprocessing is called "tightly coupled multiprocessing," because the processors:

- Share a single memory
- Are controlled by one operating system
- Reside on a high-speed internal bus

A VAX can be configured initially with a single processor (it is then called a uniprocessor), and additional processors can be added later as workload increases. The term multiprocessor, if unqualified, usually means a tightly coupled multiprocessor.

VAXclusters

DEC has developed extensions to its VMS operating system and specialized hardware components that allow both the MicroVAXes and VAXes (but not the DECstations) to be combined into VAXclusters. VAXclusters are a form of loosely coupled multiprocessing. Each computer has its own memory and operating system, and the processors are interconnected by an external, rather than an internal, bus.

The cluster architecture has these advantages:

1. Load Balancing. A cluster appears to the users to be a single system. For example, when a user logs in to a cluster through a terminal server, the session will be established transparently on whichever of the computers is least utilized.

2. Availability. If one of the computers in a cluster fails, the others can absorb its workload. The cluster can continue to operate, although performance might be degraded somewhat.

3. Optimized Data Sharing. Specialized mass storage devices can provide very highly optimized data sharing by multiple processors–not as fast as tightly coupled multiprocessing,
but faster and more efficient than, for example, using DECnet for file sharing.

4. Transparent Peripheral Sharing. To users of the cluster, peripherals attached to any
processor in the cluster appear to be locally attached. Sharing is therefore easy and efficient.

5. Scalability. As cluster workloads increase, additional processors can easily be added.

Local Area VAXcluster (LAVC)

LAVCs use an Ethernet as the external bus for the cluster. An LAVC consists of two to ninety-
six processors, typically MicroVAXes. Normally, one MicroVAX is equipped with large, fast
disk storage devices to be shared by all of the processors in the cluster. This node is called the
boot-node because, if any of the other processors in the LAVC are diskless, it will provide them
with a copy of the operating system (See Figure 8-2).

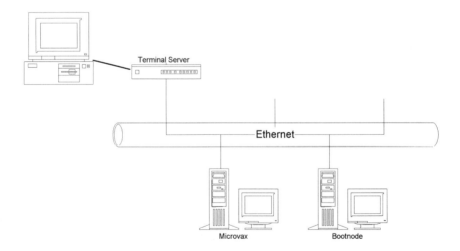

Figure 8-2. Bootnode

The Ethernet external bus operates at the standard speed of 10 Mbps. This Ethernet can be dedicated to connecting clusters or can be shared with other networks, such as DECnet. Processors in a cluster can be physically located anywhere the Ethernet will allow (See DEC Coupler). Terminal servers attach multiple terminal devices to the cluster.

A VAXcluster is an arrangement where clusters rely on a specialized external bus and consist of:

1. From two to ninety-six VAXes, which can be multiprocessors.

2. A Computer Interconnect (CI) Bus, the high-speed external bus that interconnects the systems. It operates at 70 Mbps and consists of four coaxial cables up to forty-five feet long for each system, all joined together in a central Star Coupler.

3. A Hierarchical Storage Controller (HSC), an intelligent mass storage device consisting of large, fast disks, tape drives, and a specialized processor. It connects to the CI bus, servicing I/O requests for the processors in the cluster. Two HSCs can share disk drives, backing up one another and optionally providing shadowing of critical data by automatically keeping an up-to-date copy of it.

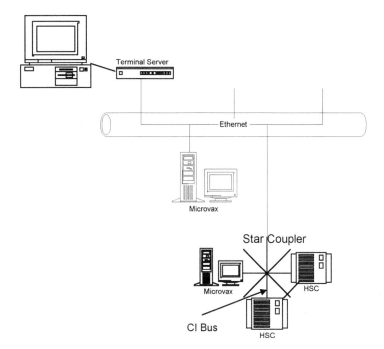

Figure 8-3. VAXcluster

4. Terminal servers connected via Ethernet.

5. A System Communication Architecture (SCA). This specialized communication architecture is used by the MicroVAXes and VAXes in a cluster to communicate across the Ethernet or CI bus. SCA is an optimized, two-layer architecture for high throughput, used only for this purpose. In the case of the LAVC, the same Ethernet channel can be used simultaneously for SNA, DNA and LAT protocols.

DECnet / SNA Gateway

DEC sells a variety of products which can act as DECnet/SNA gateways (See Figure 8-4). A VAX running SNA gateway software called VMS/SNA and equipped with an IBM channel adapter can connect directly to an IBM mainframe channel. Alternatively, three special purpose gateway devices which can connect to an IBM channel or a 37x5 FEP are available.

Although workstations make up an increasing portion of DEC's business, multiuser systems is a more traditional DEC environment. The terminals are character-oriented. Each character that the user types must be "echoed" by the host computer in order for the user to see it on the screen.

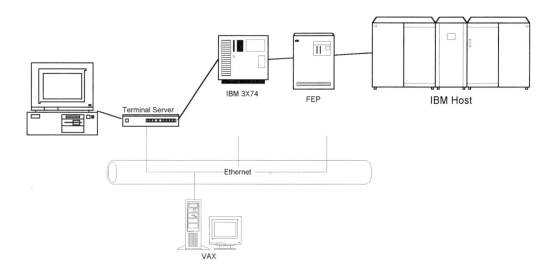

Figure 8-4. SNA Gateway

Terminal Servers and LAT

Terminals can be directly attached to the host, or they can reside on an Ethernet network. DEC has selected Ethernet as its standard for Local Area Networks. Today, many "terminals" are actually PCs running a terminal emulator program. Whether attached directly or through Ethernet, the host must interrupt application processing twice for each character typed on the terminal. This results in significant overhead for the host, with cycles that could be spent on application computing going instead to low-level communications tasks.

To address this problem, DEC developed terminal servers, called DECservers, to connect terminals to VAXes or clusters, and the Local Area Transport (LAT) architecture. LAT is implemented by the DECservers and the VAXes or clusters to transfer character data via Ethernet (See Figure 8-5). In LAT terminology, the VAXes and clusters are services. The term LAT also refers to the protocol used by nodes which support the architecture.

A terminal server can provide up to 128 ports (model dependent). Each port can connect a local or remote terminal (or a PC which is emulating a terminal) to any of the services on the Ethernet.

One important purpose of the LAT architecture is to combine characters from several users into one Ethernet frame. This fills the frame and utilizes the Ethernet connection more fully. More important, it allows the host to handle input more efficiently, since the host can process several characters from several users each time it is interrupted for an Ethernet frame.

Another important purpose of the LAT architecture is to make services anywhere on the Ethernet available to users anywhere on the Ethernet. The DECserver can connect you to any service on the Ethernet which supports the LAT protocol. All LAT services on the Ethernet broadcast "advertisements" of their service every sixty seconds. Printers and non-LAT hosts can also be attached to the terminal server ports. For them, the role of LAT is reversed; LAT advertises

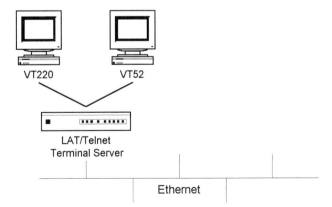

Figure 8-5. Local Area Transport

its services on the Ethernet, and any of the hosts on the Ethernet can establish a session with the service that supports the LAT protocol. For example, a host can print documents on the printer or send messages to the non-LAT host.

When one of the ports on a terminal server is connected to a non-LAT host, users of the terminal server can access that host through that port by requesting to be connected to that port, rather than to a service on the Ethernet.

DEC sells a special feature for its DECserver 510 and DECserver 550 models which turns them into "32703270 Terminal Servers." It gives them special capabilities for a mixed DEC/IBM environment. A terminal server equipped with the feature can be attached to both an Ethernet and a cluster controller. To services on the Ethernet, the 3270s appear to be DEC terminals. To the cluster controller, they look like normal 3270s. The user issues a connect command, similar to that shown above, to choose between an Ethernet or cluster controller connection.

Like SCA, LAT is a highly specialized and optimized protocol. It uses Ethernet to provide a connection to services, and then manages the flow of data between the service and the user.

Maintenance Operations Procedures

Not all nodes on a DECnet have disk drives. Desktop workstations and network nodes such as routers and terminal servers are often diskless, and thus have nowhere to store their operating system. The operating system must therefore be downloaded from another node on the network. But how can a node access the network without the operating system? The answer is MOP. Although termed a protocol, MOP is more properly a very simple architecture which implements a protocol on top of Ethernet to download the operating system. Unlike the operating system, MOP is simple enough to be implemented with read-only memory (ROM) in the node, along with the Ethernet driver. Terminal Servers, for example, download their control software from a host using MOP.

DNA OVERVIEW

DNA Version IV is already fairly close to OSI, and Version V is OSI compliant, except that the user can opt for the DNA upper layers in order to maintain compatibility with existing applications. We have combined the two diagrams used in the OSI introductory chapter to illustrate again the correspondence between DNA Version IV, DNA Version V, and OSI in Figure 8-6. Version V supports TCP/IP, which was not in DEC's original plans for the release.

The differences between the Session Control, End Communications, and Routing layers of DNA and the Session, Transport, and Network Layers of OSI are very technical. The concepts implemented by both are very similar. Therefore, this discussion will focus on more pronounced differences in their data link protocols and on several network applications.

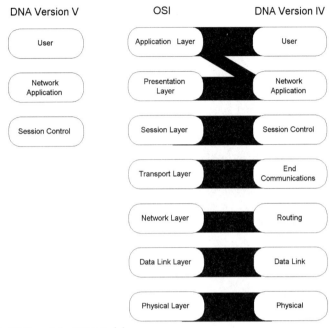

Figure 8-6. DNA and the OSI Model

Figure 8-7 illustrates the protocols we will discuss. You are already familiar with Ethernet and X.25, so we will not discuss those. We show them here to make the diagram complete.

You learned about the CI bus in the discussion of clusters in the previous section, where we said that SCA uses the CI bus for communication within the cluster. DECnet also supports the CI bus to allow sessions to take place between nodes in a cluster.

Digital Data Communications Message Protocol (DDCMP) is DEC's point-to-point data link protocol. It is similar to IBM's SDLC and OSI's HDLC. DDCMP has two important features:

1. Both asynchronous (byte oriented) and synchronous (frame-oriented) service is provided.
2. In addition to normal serial operation over a pair of wires, DDCMP will operate over multiple wires, sending the bits of a byte in parallel.

Network Applications

Communications Terminal Protocol (CTERM)

DNA's virtual terminal (VT) service for VMS systems employs this protocol. CTERM is conceptually similar to the TELNET protocol you learned about when you studied TCP/IP. CTERM is an alternative to LAT. It can be used over any supported link, whereas LAT is only available for Ethernet.

Distributed Naming Service (DNS).

This service used by VMS systems is somewhat like the DOMAIN service you learned about when you studied TCP/IP. While DOMAIN is used strictly for converting hierarchical names to Internet addresses, DNS is more general. For example, it is used to make user names and network addresses globally available for e-mail. It can also be used to make the location of files known throughout the network. With DNA Version V, DNS will automatically make a new node's name and network address available to the whole network, replacing manually maintained tables in each node.

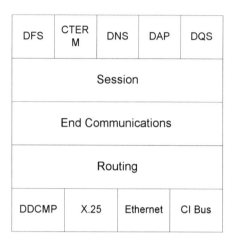

DFS	CTER M	DNS	DAP	DQS
Session				
End Communications				
Routing				
DDCMP	X.25	Ethernet		CI Bus

Figure 8-7. DNA Protocols

Data Access Protocols (DAP)

This is an older set of protocols used by VMS systems to provide remote file access to users over DNA networks. It is a client/server application which lets users access file directories and initiate file transfers through the file management systems on remote nodes.

Files can be transferred between a remote node and a local node, or between a pair of remote nodes. DAP is being gradually replaced by DFS and the OSI File Transfer, Access, and Management (FTAM) protocol.

Distributed File System (DFS)

DFS is similar to the Network File System. Like NFS, directories are identified to the network to be shared. Those directories can be mounted on remote nodes and effectively made part of that file system, just as with NFS.

Distributed Queuing Service (DQS)

This service maintains queues of "jobs" for VMS systems across the network. It is used primarily for printing, although it is a generalized service. DQS accepts jobs from applications, puts them into queues, and transfers items from queue to queue, possibly across the network. An operating system process called JOB-CONTROL eventually removes each job from its destination queue and passes it to a process called a symbiont, which is associated with the queue. The symbiont then processes the job. Standard symbionts are provided for printers. Users can create their own symbionts to handle their own queue or provide special handling for standard queues.

SUMMARY

DEC products employ two different processor architectures: Virtual Address Extension (VAX), which is proprietary, and a RISC architecture based on one developed by MIPS Computer Systems, Inc. Systems based on board-level computers are called VAXes; microprocessor-based VAX systems are called MicroVAXes. VAXstations are MicroVAX workstations. DECstations are MIPS-based workstations. VAXes and MicroVAXes run the Virtual Memory System (VMS) operating system and ULTRIX, DEC's version of 4.3 BSD UNIX. DECstations run only ULTRIX.

Recent models of VAXes feature tightly coupled multiprocessing, where multiple CPUs reside on a high-speed internal bus, share a single memory, and are controlled by one operating system.

Clusters are loosely coupled multiprocessors, connected by an external bus. Each has its own memory and operating system. A Local Area VAXcluster (LAVC) interconnects up to ninety-six MicroVAXes or VAXes and terminal servers on an Ethernet channel. A VAXcluster combines up to ninety-six VAXes on a Computer Interconnect (CI) bus, which operates at 70 Mbps. A VAXcluster includes one or more Hierarchical Storage Controllers (HSCs), which provide fast access across the CI bus to mass storage devices (disk and tape), and can include one or more terminal servers.

Processors in a cluster use the System Communication Architecture (SCA) to communicate across the Ethernet or CI bus. SCA is a highly specialized, two-layer architecture.

Clusters provide load balancing, improved availability, optimized data sharing, transparent resource sharing, and scalability.

DECserver terminal servers implement the Local Area Transport (LAT) architecture to combine characters from several terminals into a single Ethernet frame, making more efficient use of the Ethernet channel and, more important, reducing the host overhead associated with processing separate messages for each character typed by a terminal user. A DECserver can provide up to 128 ports for local or remote terminals. Nodes on the Ethernet that support the LAT architecture advertise their services on the Ethernet once every sixty seconds. The terminal servers make advertised services available to users of the ports.

The Maintenance Operations Protocol (MOP), actually an architecture rather than a protocol, downloads operating systems to nodes and terminal servers which do not have their own local disk storage.

Version IV of DNA is already close to OSI, and Version V is OSI compliant, except that the user may opt for the DNA Session Control, Network Application, and User Layers for compatibility with Version IV applications. DEC is committed to delivering DECnet, OSI, and TCP/IP on all platforms, but their strategy for migration from DNA Version IV to DNA Version V was a difficult one.

The most noticeable differences between OSI and DNA are at the extremes of the functional layers, the Data Link Layer and the Application Layer. The Data Link Layer includes two protocols in addition to those found in the OSI Data Link Layer: Digital Data Communications Message Protocol (DDCMP) and CI bus. DDCMP is DEC's equivalent to SDLC or HDLC, but supports both asynchronous and synchronous communications. The CI bus is supported to enable nodes within a cluster to communicate.

Network applications include:

Communications Terminal Protocol (CTERM). DNA's virtual terminal (VT) service for VMS systems employs this protocol, which is an alternative to LAT that can be used on any link.

Distributed Naming Service (DNS). DNS is a generalized naming service for VMS systems, allowing the names of users, files, nodes, etc., to be known to the whole network.

Data Access Protocols (DAP). DAP is an older set of protocols which provide remote file access for VMS systems. DAP is being supplanted by DFS.

Distributed File System (DFS). Similar to the Network File System, directories are identified to the network to be shared and then can be mounted on remote nodes.

Distributed Queuing Service (DQS). Used primarily for printing, this service for VMS systems maintains queues of "jobs" across the network. JOB CONTROL removes jobs from queues and passes them to symbionts, which process the job.

9

INTERNETWORKING OF ARCHITECTURES

INTRODUCTION TO INTERNETWORKING

This chapter introduces you to the topic of internetworking. In Chapter 12 we look at an actual network and the components that make up a typical network. When considering internetworking, two issues should be considered. The first is the division of a network into manageable pieces. As networks grow, certain problems arise and management of the network becomes increasingly difficult to manage. This chapter will discuss the problems that are encountered with network growth and the internetworking components that solve these problems.

The second issue comes when the need arises to connect separate networks. When corporations merge or divisions of a company want to share information, problems arise. This chapter will explain problems associated with this type of growth including problems found when trying to connect networks that are different architectures.

CHAPTER OBJECTIVES

- Identify internetworking components such as repeaters, bridges, routers, switches, hubs, and gateways

- Identify the layer of the OSI model which corresponds to the different types of internetworking components

- Identify the types and functions of the various bridging and routing methods

- Understand when to bridge, route, or switch

- Understand the functions and uses of a TCP/IP router

- Understand what takes place when routing data over an internet

- Identify the major problems that can be solved using a protocol converter

- Understand network management issues and concerns

Integration of Networks

As LANs grow and as clients and servers are added, traffic congestion becomes a problem. Figure 9-1 demonstrates the results of many clients accessing the same physical media for usage of networked devices such as print servers and file servers.

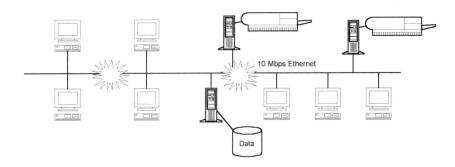

Figure 9-1. LAN Growth

Figure 9-2 demonstrates the situation that is sometimes found when separate companies or different departments of one company need to share information. Two companies, one in Denver and another in Atlanta, merge and need to share information. The company in Atlanta uses a Novell NetWare network and the company in Denver uses an IBM host. Many questions emerge, including:

- How do they share information from different data formats?
- How do they connect two different types of architectures together?
- What is the best solution for tying the two networks together over a wide area?

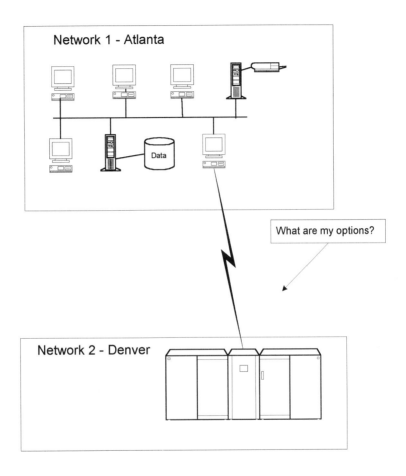

Figure 9-2. Network Consolidation

RELAYS AND REPEATERS

A relay is any device or node that connects two subnets or networks. Repeaters are the simplest type of relay or internetworking device. A repeater operates at the Physical Layer of the OSI model (See Figure 9-3). Since the physical layer is concerned with bits, the job of a repeater is to repeat bits. If a "1" bit is received on the input port of a repeater, a "1" bit is regenerated at the output of the repeater. If a "0" bit is received on the input port of a repeater, a "0" bit is regenerated at the output of the repeater. All information received is passed on to each connected segment. A repeater is considered a "nondiscriminating" device.

Figure 9-3. Repeaters and the OSI Model

As nodes are added to a LAN, the cable can be extended with electrically passive devices that simply connect one segment to another. But eventually the permissible length of the cable segment will be exceeded, or it will become necessary to connect different cable types, or it might be desirable to run more than two cables outward from a central point. A repeater is used to take care of these needs.

A repeater connects one segment of a LAN to another, possibly connecting differing types of media. For example, a repeater can connect thin Ethernet cables to thick Ethernet cables as shown in Figure 9-4. It regenerates electrical signals from one segment of cable to another. Since it reproduces exactly what it receives, bit by bit, it also reproduces errors. But it is very fast (10Mbps for Ethernet), and causes very little delay.

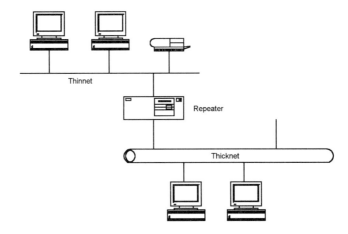

Figure 9-4. Repeater and Media Types

A repeater cannot connect two different media access types (MAC Frames) such as Token Ring and Ethernet. Keep in mind that a repeater is a physical layer device and does not care where a frame starts and stops or what the contents or format of a frame might be.

As internetworking devices for Ethernet LANs, repeaters are feasible only for relatively small LANs (less than 100 nodes) which are confined to a small geographical area, such as one or two floors of an office building. A repeater should not be used to connect heavily used LANs together, since a repeater cannot isolate traffic between LAN segments. Since each bit is copied to the attached segments, all data will pass through a repeater. Therefore, if you connect multiple LAN segments together using a repeater, you may experience performance problems, since repeaters do not filter out any data passing through them.

Examples of Repeaters

Ethernet Repeaters

In an Ethernet LAN, a repeater is used to extend the length of the physical media. Three Ethernet Repeaters can connect three Ethernet 10Base5 segments together using two repeaters. In a configuration such as this, repeaters are considered to be a node on EACH segment. Therefore the maximum number of nodes that can be attached to this type of network is 296.

Repeaters can be used to attach "link segments" to extend the overall distance of a network. This is shown in Figure 9-5. Note that two of the five segments do not contain nodes. This is an example of the 5/4/3 rule. You can connect a maximum of 5 segments together using 4 repeaters, but only three of the segments can contain devices (other than repeaters).

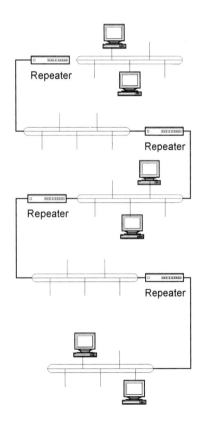

Figure 9-5. Ethernet Repeaters and Link Segments

Figure 9-6 shows distance considerations for repeaters.

Media Type	Connector Type	Max Segment Length	Max Nodes	Max Nodes Per Segment	Minimum Distance Between Nodes	Combined Length of all Segments
10Base5	DB-15	500 Meters	300 including repeaters	100 including repeaters	2.5 Meters 8 Feet	2500 Meters 8200 Feet
10Base2	BNC	185 Meters	90 including repeaters	30 including repaters	.5 Meters 1.5 Feet	925 Meters 3035 Feet
10BaseT	RJ-45	100 Meters to Hub	4 Linked Hubs	Hub dependent	n/a	2500 Meters 8200 Feet

Figure 9-6. Ethernet Repeater Specifications

Token Ring Repeaters

In a Token Ring LAN, there are three different types of repeaters. The first type is found at each individual node typically in the Network Interface Card (See Figure 9-7).

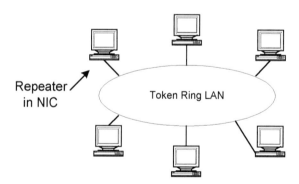

Figure 9-7. Token Ring NIC Repeaters

The second type of repeater found in a Token Ring network is referred to as the "lobe repeater" (Token Ring nodes are referred to as lobes). A lobe repeater allows Token Ring nodes to be located further from the MAU then normal configurations as shown in Figure 9-8.

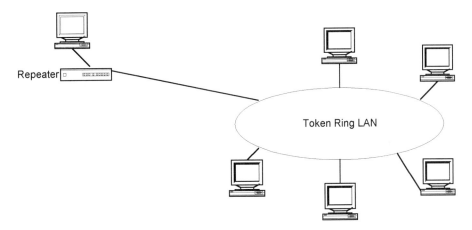

Figure 9-8. Lobe Repeaters

The third type of Token Ring repeater is shown in Figure 9-9. This type of repeater is used when there are multiple MAUs on the network. The IBM 8218 is an example of a ring repeater. It can extend the ring path up to 750 meters in length.

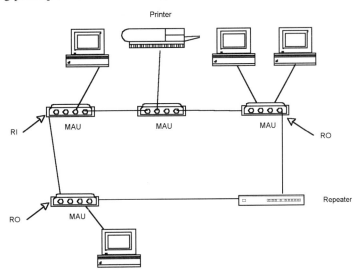

Figure 9-9. Ring Repeater

ESCON Repeaters

An ESCON communication system consists of a number of different components, one of which is an ESCON repeater (9034) (See Figure 9-10). The ESCON repeater regenerates bi-directional ESCON signals on full-duplex ESCON links to extend the distance between an ESCON channel and control unit or another ESCON channel. Channel to channel connections as long as 60 km are possible with two chained repeaters on a single-mode link. Connections as long as 9 km are possible using two chained repeaters on multimode links. The maximum extension distance depends on devices such as directors and types of controllers installed in the system and fiber cable quality.

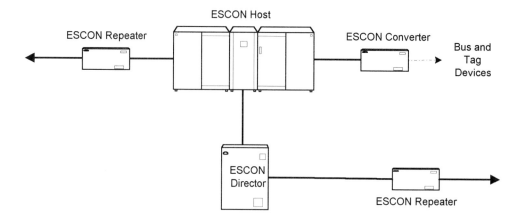

Figure 9-10. ESCON Repeater

BRIDGES

Bridges operate at the Data Link Layer of the OSI model and are sometimes called "level 2 devices," or alternatively "link layer devices" (See Figure 9-11). They are completely transparent to the network and upper layers and therefore can accommodate more than one communications architecture. They operate on frames and have no regard for the content of the frame. For example, a bridge could link LANs with nodes that use both the TCP/IP and DECnet architectures. This is not necessarily true for routers that operate at a higher level, as we will see later in the chapter on routers.

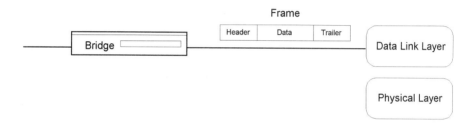

Figure 9-11. Bridges and the OSI Model

Bridges are more complex than repeaters. Repeaters are primarily hardware devices which repeat signals from segment to segment and typically do not contain operational software. The operation of bridges requires both hardware and software. Bridges perform functions such as filtering and forwarding MAC frames. A bridge determines if the incoming frame is destined for a device which is on the segment where the frame was generated. If it is, the bridge does not forward the frame to the other bridge ports. This is an example of filtering. If the MAC destination address is on another segment, the bridges sends the frame to the appropriate segment. This is known as forwarding. Figure 9-12 illustrates bridge ports and bridge tables. When a frame comes in with a destination address of LAN1, the bridge will forward the frame to the segment attached to port 3. If a frame is received with a destination address of LAN3, the bridge will forward the frame to the segment attached to port 1.

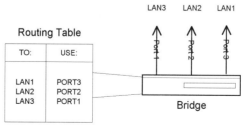

Figure 9-12. Frame Forwarding

Bridges come in many shapes and sizes. The simplest of bridges can be an adapter card (NIC) in a PC which attaches to small LAN segments. The most elaborate bridges convert MAC frames from one type to another and/or route frames over long distances at very high speeds.

Bridges and Traffic Isolation

In both Ethernet and Token Ring LANs, the LAN is busy whenever any node is transmitting. As the number of nodes on the LAN increases, so does the utilization of the LAN. At some point, the LAN will become a bottleneck; network users will have degraded performance and will find themselves spending too much time waiting to access the LAN. However, traffic on a LAN tends to be local. Nodes within a department tend to communicate with one another much more than they do with nodes in other departments. Frequently, a "server" node provides a virtual filestore for a department, and much of the traffic tends to be between user nodes and the server. If a LAN is divided into two subnets, the utilization of each subnet will be approximately one-half what it was before. A bridge can then be used to interconnect the two parts, and a relatively small proportion of network traffic will need to flow across the bridge. The effect of the division of the network is called "traffic isolation," and it is a primary tool for controlling network utilization (See Figure 9-13).

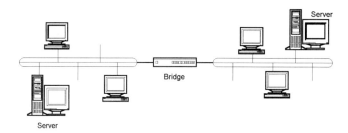

Figure 9-13. Bridges and Traffic Isolation

Bridges and Format Conversion

In many businesses today, computer networks exist which were installed to serve varying needs in different departments. These systems were often designed to be purely internal to a group, and no attempt was made to be consistent with other hardware or software already in use at the time. As employees become comfortable with the ability to share information on a network and companies grow and become more automated, a need to share information and resources between these diverse networks may result. One common situation is the presence of personal computers on Token Ring LANs and UNIX based workstations or mainframes on Ethernet LANs (See Figure 9-14). With the right combination of hardware, such as a bridge and software, the resources of these diverse systems can be made readily available to all of the users of both systems. This is a good example of the need to "interoperate."

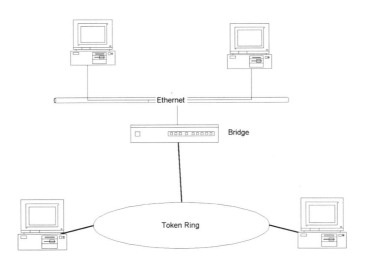

Figure 9-14. Ethernet and Token Ring Bridging

If we analyzed the protocol stacks, one thing will become apparent: bridges do not manipulate the information inside of the frame (See Figure 9-15). Therefore, for two devices to communicate across a bridge, the upper layer protocols must be the same or another device, such as a protocol converter, must be introduced into the network.

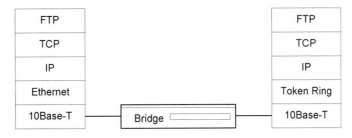

Figure 9-15. Bridges and Protocol Stacks

Wide Area Bridges

As an enterprise network grows, it will tend to be dispersed over a wide geographical area. Another important use of bridges is the ability to connect LANs remotely (See Figure 9-16). WAN bridges, also called half-bridges, work together in pairs, as shown. One port of each bridge is connected to a leased line or, alternatively, to an X.25 network. The bridges cooperate to route frames across the point-to-point link using a protocol such as SDLC or HDLC, or across the public network using X.25. The point-to-point link, such as 56.6 Kbps, is much slower than either Token Ring or Ethernet, so traffic between the two LAN segments must be relatively low.

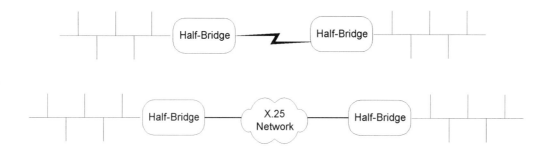

Figure 9-16. Wide Area Bridges

Token Ring Bridges

Figure 9-17 shows how Token ring networks can be joined using bridges. The purpose for using a bridge, as mentioned earlier, is usually to isolate local traffic on one ring from local traffic on another ring. The rings should be arranged so that most of the traffic on one segment of the network stays on its own ring rather than crossing the bridge. Adding a second bridge would provide redundancy and higher levels of availability.

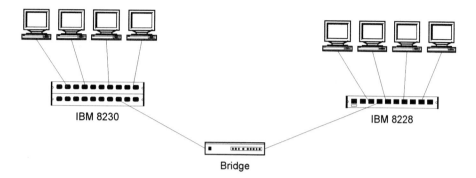

Figure 9-17. Token Ring Bridge

Figure 9-18 shows how to form a backbone configuration using Token Ring networks and bridges. This configuration allows for any-to-any communication between network nodes across several rings. When networks are large enough to have multiple rings, the backbone configuration provides the shortest average path between any two nodes on the network. Shared devices such as printers and file servers can be placed on the backbone to provide quick access to all nodes on the network. A typical implementation for this type of configuration is the connection of multiple floors in a multistory office complex. Each floor may contain a separate ring and all rings (floors) can be connected via the backbone.

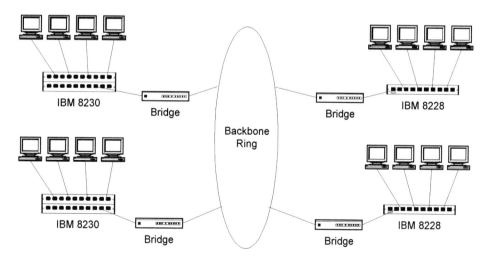

Figure 9-18 Token Ring Backbone Connectivity

In many instances, fiber optics (such as Token Ring or FDDI) are used as the backbone. This eliminates problems associated with differing ground potentials found in separate buildings and environmental hazards. Figure 9-19 shows how Token Ring networks can be connected to an FDDI backbone. This is a very typical configuration in today's LAN environment. Token Ring and Ethernet LAN segments connect to high-speed backbones via bridges or routers for connectivity to other segments or across Wide Area Networks (WANs).

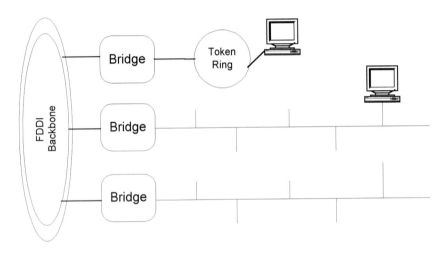

Figure 9-19. Bridges and FDDI Backbones

Ethernet Bridges

Figure 9-20 shows an Ethernet bridge configuration consisting of two individual Ethernet segments. Bridge filtering keeps traffic on local segments unless the frames are destined for another segment or contain broadcast addresses.

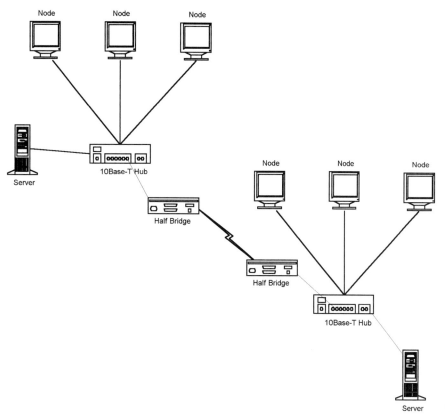

Figure 9-20. Ethernet Bridge

Bridging Algorithms

Looking at the issues of bridges from a broad perspective, two important issues should be noted (See Figure 9-21):

	Fixed	Adaptive
Connectionless	Fixed tables in bridges	IEEE Spanning Tree Bridges 802.1
Connection-oriented	Fixed tables in source node	IEEE Source Routing Bridges

Figure 9-21. Bridge Types

- Whether bridging is connectionless or connection-oriented.
- Whether the method is fixed or static, requiring changes by a network administrator, or adaptive, which automatically accommodates changes in the network.

IEEE Spanning Tree Bridges are connectionless because the routing is done by the bridges. Each frame is routed independently. These bridges are adaptive because the tables used for routing are created dynamically and will quickly and transparently be adapted to changes in the network. This is why they are sometimes called transparent bridges.

IEEE Source Routing Bridges are connection-oriented because the routing is done by the source node. Once a route is established, it will not vary, which suits a connection-oriented system. They are adaptive because the tables used for routing are created dynamically and will adapt to changes in the network. However, because of the way the tables are constructed and because of the very nature of a connection-oriented facility, source routing bridges are not nearly as "transparent" as their spanning tree counterparts.

To understand this discussion of bridges, you need to remember two points:

1. Every node in the LAN sees every frame transmitted on its own LAN. In the case of a LAN bridge, the bridge sees every frame transmitted on all of the subnets to which it is attached.

2. A bridge always knows the subnet or network ID of those LANs to which it is attached, because it must know its own addresses, which contain the IDs. But it does not know about any other bridges attached to those subnets or networks.

Broadcasting

Since both types of bridges we will study are adaptive, they must have some method of discovering the routes that lead from one node to another in the network. Broadcasting plays an important part in route discovery for both types of bridges. We will first discuss broadcasting in general. Later in this section, you will see how it is used in the two types of bridges.

Broadcasting is also called flooding because, as the name implies, a node floods the network with frames. It is also called forwarding. When a frame is broadcast, its header contains a flag, indicating this to the bridges in the network. Seeing this flag, each bridge simply retransmits the frame on each of its ports except the one on which it was received.

Therefore, a node can communicate with every other node in the entire Internet by transmitting a frame with the appropriate header flag set. However, broadcasting is susceptible to "looping" of frames in networks where bridges create closed circuits. Consider Figure 9-22. If node A on LAN L1 broadcasts a frame, Bridge 1 (b1) sends it to b2, who sends it to b3, who sends it back to b2, who sends it to b3 . . . and so on, unless something is done to prevent it.

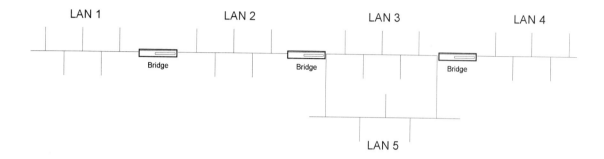

Figure 9-22. Bridges and Broadcasting

The Spanning Tree

A spanning tree is a logical tree structure that "spans" a network so that there are no loops (See Figure 9-23). On the top of this diagram you see a network of bridged LANs. We are showing it in this different arrangement to illustrate the "tree." The selection of a bridge to be the root of a spanning tree is more or less arbitrary and is taken care of automatically by the bridges themselves. This diagram could be redrawn with any of the bridges as roots.

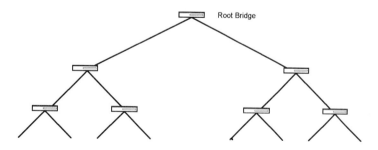

Figure 9-23. The Spanning Tree Structure

We can make several observations about the effect of imposing the spanning tree on the network:

- There is at most one path on the tree between two nodes; if this were not the case, there would be a potential loop.

- When a frame is broadcast, each subnet receives at most one copy of the frame.

- When a frame is broadcast, only as many copies are made as there are subnets.

As you see, the creation of the spanning tree eliminates some possibilities for parallel data flow and causes some less than optimal routing of network traffic. This is a potential cost that must be paid for eliminating the looping of broadcast frames.

The effects of the spanning tree might need to be taken into account by the Network Administrator when installing bridges which create parallel paths. Those bridges might not always have the desired effect.

Note that the concept of "spanning tree" is employed by both spanning tree bridges and source routing bridges, but in somewhat different ways. Whereas the spanning tree is used all of the time in spanning tree network, it is used only intermittently in a source routing environment, where it is used only to send discovery frames.

Transparent (Spanning Tree) Bridges

Because transparent bridges are meant to be connectionless, the routing is done by the bridges. When a bridge is activated, it contains no routing information. This poses two problems:

- The routing table must be created.

- Each bridge must somehow handle the routing of frames without a routing table until the table is created.

Creating the Routing Tables

Routing tables are created through a process called "backward learning." Figure 9-24 illustrates the process. The bridge simply watches traffic on the network. When it sees a frame go by on one of its ports, it knows that the subnet that originated the frame is "reachable" from that port. It puts that information into its routing table. In the illustration, it sees a frame on Port 1 that came from LAN 1 (L1), so it knows it can reach L1 on Port 1. Once an entry is made in the table, and the bridge receives a frame addressed to that subnet, it knows on which of its ports it should transmit it.

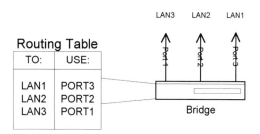

Figure 9-24. Bridge Tables

In this way, the bridge quickly establishes routing information for all subnets which have nodes that are currently transmitting frames. Entries in the routing table are not permanent, however. The bridge periodically purges all entries in its table which are more than a few seconds old. Thus, when a failure occurs somewhere on the network, or when the configuration of the network changes and a subnet is no longer reachable on a given port, the bridge will not be receiving frames from the subnet on that port, and the table entry will soon be purged. Then, new entries will be made in the table to correspond with the new configuration of the network.

Broadcast Routing (Flooding)

But what happens when a bridge receives a frame addressed to a subnet not in the bridge's table? This will happen while the bridge is initially establishing its routing table, when an inactive subnet becomes active, and when a bridge is inserted into the network. This is where the spanning tree comes in.

The bridge broadcasts the frame, "flooding" the network with copies. It broadcasts only on the spanning tree, so frames cannot loop.

Other bridges in the network retransmit the broadcast frame on all other subnets to which they are attached and which are on the spanning tree (unless they are attached to the target subnet, in which case they will retransmit the frame only on that subnet).

The frame will reach its target node eventually. It usually will be acknowledged. When the source bridge sees the acknowledgment frame, it learns the route to that subnet and will not have to flood any more frames addressed to it.

Now you can see the "transparent" nature of the spanning tree bridge. Within seconds of a change in the network, a new spanning tree has been calculated, and all of the tables in all of the bridges on the network reflect the change in the network. It will be momentarily necessary for some frames to be flooded, but they will be confined to the new spanning tree. Therefore, not too many additional frames are generated, and each subnet receives at most one copy.

You can also see the lack of connectedness. A node can send two frames and, because of changes in the network during their transmission, they can follow different routes—something which can never happen in a connection-oriented system.

Source Routing

For source routing, routing is done by the nodes in the network rather than the bridges. Each node must maintain its own routing table for all its connections. The table can be small, because there are only a few connections for each node and, for each connection, usually no more than five or six hops. Source routing provides an automated way of establishing these tables in the nodes.

Each node begins with an empty routing table. These are the steps taken to create the routing table:

- The first time a node needs to establish a connection, it sends a discovery frame addressed to the target node. As mentioned earlier, source routing also uses the concept of a spanning tree. Specifically, the frame is flagged to be flooded on the spanning tree.

- The target node receives the discovery frame. Because the spanning tree is used for flooding, the target node will receive only one discovery frame.

- The target node answers with a response frame. The response frame is flagged to be flooded to the entire Internet, not just on the spanning tree.

- As each bridge in the network relays the response frame, it records its portion of the path taken by the frame in the frame header. Looping is avoided because the bridge will check the route in the header to see if the frame has already passed by, and if so, discard it.

- The source node receives a response frame for every possible path between it and the target node. The complete path taken by the frame is recorded in its header.

- The source node chooses the optimal path from the answer frames it receives based on number of hops, and stores that in its table.

- The path stored in the table will be used for the entire life of the connection and for subsequent connections, barring problems. Each frame that is transmitted to the target node will contain the explicit routing from the table in the source node. Source routing bridges thus perform three functions:

    1. Maintain the spanning tree for the network

    2. Relay discovery and response frames on all subnets to which the bridge is attached (except the one on which the frame was received)

    3. Relay other frames according to the route stored by the source node in the frame header

What happens when something in the network changes? Suppose a bridge goes down? Since routing is contained in the frame, the bridges can do nothing but discard a frame that is undeliverable (that is, their retransmission of the frame was not acknowledged). Ultimately, the source node determines that the target is not responding, because it hasn't received an acknowledgment. It then must reestablish the connection by sending discovery frames as it did before. Meanwhile, the bridges will have rebuilt their spanning tree table to accommodate the change in the network, and the rediscovery of the route by the source node will result in a different route to the target node, if one exists. If no alternate route exists, the source node will not receive a response to its discovery frame within the allotted time and will assume that the node is unreachable.

Bridge Limitations

Bridges are fast and adaptive, making them relatively easy to install and use compared to internetworking components that operate at a higher level. But they do have problems, especially in relation to their use of broadcast frames. A consequence of broadcast traffic is that problems that arise between a pair of machines can be seen by the entire network.

Broadcast Storms

Due to differences between nodes and bridges in different parts of the network, a broadcast frame can sometimes be misinterpreted. This leads to another broadcast frame by the bridge which misinterprets it. The second broadcast frame is again misinterpreted, and so on. The result is a "storm" of broadcast frames which can severely impact network performance. Sometimes storms can persist and eventually bring down the entire network. Such problems can be very difficult to resolve.

Broadcast Traffic Growth

As bridges are added to a network, the growth in broadcast traffic is not linear but exponential. Eventually, a significant portion of the network bandwidth can be used by broadcast frames.

Poison Packets

When changes are made to network software, program problems can occur. Sometimes they will occur in such a way that the receipt of a broadcast frame of a certain type can cause a catastrophic problem in each bridge or node which receives it. These are called "poison packets" because when broadcast, they can cause users all over the network to experience problems all at the same time.

Interconnnectivity Limitations

Bridges are also limited for internetworking in heterogeneous environments because they cannot interconnect networks which have different Data Link Layer protocols (other than bridging between IEEE 802 protocols, as mentioned). For example, a bridge cannot connect a LAN to a point-to-point WAN.

These limitations are overcome by using internetworking devices which operate at a higher level in the OSI stack, such as routers and gateways.

Numerous proprietary bridge protocols have been implemented in a variety of bridge products, but standards for bridge protocols are emerging. As you have seen, the important Data Link Layer protocols for LANs are IEEE 802.3 Ethernet and IEEE 802.5 Token Ring. Each has a corresponding IEEE 802 draft standard for bridge routing:

- The IEEE 802.3-related protocol is IEEE 802.1, "Spanning tree." To the extent routing is standardized, this is the protocol used in most Ethernet networks.

- The IEEE 802.5-related protocol will be an addendum to that standard and is called "Source Routing." This is the protocol used in most Token Ring networks.

ROUTERS

Routers operate at the Network Layer rather than the Data Link Layer of the OSI model (See Figure 9-25). Routers interconnect networks which have the same communications architecture, but possibly different lower level architectures. In other words, routers are protocol dependent. A router can usually be used instead of a bridge. Previously, routers have been slower and more expensive than bridges; now, however, both price and performance of routers is approaching that of bridges. Note that in order to be "routable," an architecture must obviously have a Network

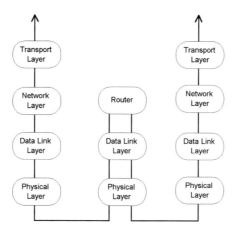

Figure 9-25. Routers and the OSI Model

Layer. Not all do. The classic example is the DEC LAT protocol you learned about in the previous chapter. It must be bridged. Since routers operate at the network layer, they operate on packets.

A router examines information located in the header of a packet and routes the packet along a path to its final destination. The network layer header contains a network address that a router uses to determine the final destination. A router looks up this destination address in its routing table and decides which link to put the packet on (See Figure 9-26).

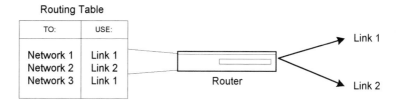

Figure 9-26. Routing Tables

Firewalls

Routers can be used to erect a "firewall" between parts of the network (See Figure 9-27). The protocol used for routing by the Network Layers of the various communication architectures do not involve the sending of broadcast packets. By using routers, as shown here, broadcast messages are kept behind a "firewall." When addressing and other problems occur, they are experienced by only a small group of users and are much easier to localize and resolve.

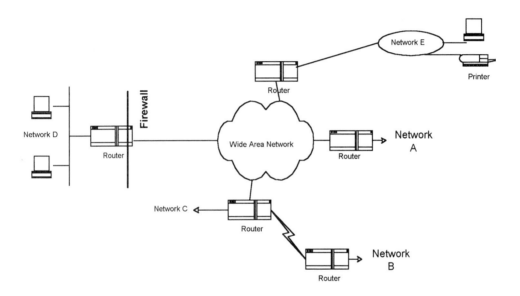

Figure 9-27. Routers and Firewalls

Multiple Data Link Protocols

Because routers operate at the Network Layer level, they can rely on the lower layers to present uniform packets, stripping off the Data Link Layer protocol headers, and combine and sequence frames as necessary. It is therefore easy for a router to handle differing Data Link Layer protocols in a generalized manner, unlike the specialized multiprotocol bridges mentioned later. For example, a router can connect LANs to a WAN, or can create a backbone network over a wide area (a WAN backbone).

Multiprotocol Routers

While routers can easily handle different Data Link Layer protocols, higher level protocols are a different matter. If two LANs are to be interconnected, and if two communication architectures—for example, DNA and TCP/IP—are used in both LANs, then two routers can be used, one for each architecture. Each will use the same set of Data Link Layer protocols, of course. Communications between TCP/IP nodes will go through one router, while communications between DNA nodes go through the other, as shown in the Multiprotocol Router diagram and the Multiple Network diagram.

To avoid the expense and additional administration of having two routers with separate LAN ports, multiprotocol routers have been developed. However, these devices do not perform protocol conversion above the Data Link Layer. They simply bundle the function of two routers into a single box and require a single port on each LAN to which they are connected.

Multiple Data Paths

Routers are also used to provide multiple paths between source and destination nodes (See Figure 9-28). This adds fault tolerance and redundancy to networks. In some cases, routers also provide flow control when networks are congested. The Alternate Route diagram shows a network configuration using routers to provide multiple paths between nodes.

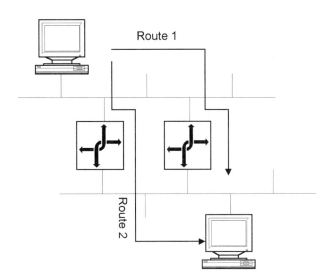

Figure 9-28. Routers and Multiple Data Paths

TCP/IP Routers

You learned in an earlier chapter how TCP/IP Internet Addresses provide an addressing scheme that is consistent within and across autonomous systems. We have also briefly discussed how messages can be routed within autonomous systems by using bridges. In this section, we will explain how messages are routed between autonomous systems (See Figure 9-29).

Routing in IP is based on the Internet addressees and on routing tables maintained in the router. For each network that the router knows about, routing tables give the Internet network address and the port on the router used to reach that address.

Each TCP/IP router that lies at the boundary of an autonomous system uses a protocol called the Exterior Gateway Protocol (EGP) to communicate with routers in other autonomous systems (routers in TCP/IP are called gateways).

The router uses EGP to advertise the network IDs of the networks within its own autonomous system to other routers. It essentially tells the other routers periodically, "Here are the networks you can reach by sending messages to me..." and the other routers reciprocate by sending back the list for their own autonomous systems.

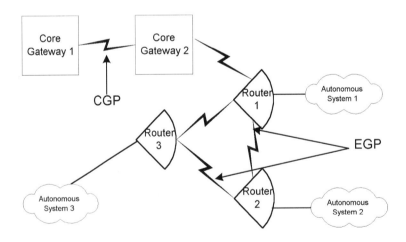

Figure 9-29. TCP/IP Router

EGP provides only "reachability" information, rather than full routing information. It tells that the network is reachable, and it gives the number of "hops" required to get through the autonomous system to the target network. But no other "cost" information, for example, on the speed of the links to the network is made available. Because the speed and cost of "hops" can vary so widely (from a 9600 baud phone link to a 100 Mbps optical fiber link), the recipient of EGP routing information cannot tell the best route to a network.

Therefore, routers use EGP to advertise their own networks only. In the diagram, Router 1 (r1) would advertise its networks to r2 and r3 and, in exchange, would learn which systems it can reach through those routers. But it would not advertise System 3 networks to r2 or system 2 networks to r3. This restriction prevents loops or unwanted paths.

However, r1 will use EGP to advertise System 2 and System 3, as well as its own networks, to a "core gateway." The core gateways are systems which form the backbone of a large TCP/IP network. In particular, several core gateway minicomputers form the backbone for the Internet.

This has the effect of mapping a spanning tree onto the networks, with the core gateway as its root. The several core gateways use another protocol, called Core Gateway Protocol, which is specialized for use by core gateways only. With it, they advertise the "reachability" of the networks they have learned about, so every core gateway contains complete reachability information for the entire Internet.

Since the gateways advertise reachability frequently, the Internet routing scheme is moderately adaptive. But, unlike the bridge protocols we've seen, it cannot adapt fast enough to handle a temporary glitch in the network.

GATEWAYS

As you have seen, bridges and routers can be used in networks that run more than one communications architecture. But they cannot interconnect nodes which use different architectures. TCP/IP nodes can communicate with other TCP/IP nodes, but not with DNA nodes. As many types of gateways are possible as there are combinations of communications architectures and application level protocols.

Protocol Converters

One type of gateway, called a protocol converter, changes a protocol from that of one communications architecture to that of another. An example of a protocol converter is a node which can connect a TCP/IP network to an SNA network, as shown in Figure 9-30. Another example is a node which converts OSI MOTIS mail to SMTP for TCP/IP delivery. A gateway handles differences in upper level protocols. To illustrate this point, we will use two combinations of protocols, SNA and TCP/IP to show gateways.

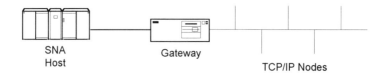

Figure 9-30. Gateways and Protocol Conversion

A protocol converter must operate at all of the protocol levels above the Data Link layer and must be transparent to the processes in those layers at each end of the connection. Figure 9-31 depicts the correspondence between TCP/IP and SNA through their mutual correspondence to OSI. It is obviously not an exact match. Two problems are apparent:

1. The Protocols do not correspond exactly and cannot be independently converted. Instead their conversion must be handled as an integrated task. For example, when converting from IP to SNA Path Control protocol, elements of the IP protocol must also be reflected in the SNA Transmission Control Protocol.

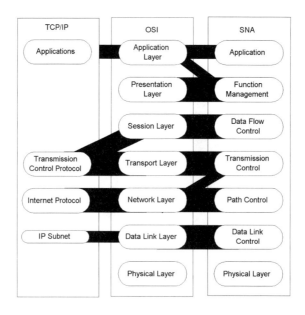

Figure 9-31. TCP/IP and SNA Architecture Comparison

2. Also note that layers are missing. For example, the conversion program must handle the Function Management protocol even though that layer does not exist on the TCP/IP side.

The task of protocol conversion can also be viewed in relation to the datastream (See Figure 9-32). The converter sees datagrams from the TCP/IP side and BIU from the SNA side. A conversion of header data must be performed by the conversion program, i.e. the datagram header must be removed and headers for the BLU, BIU, and RU with the corresponding information must be built. Note that protocol converters must also perform the function of a router, maintaining consistent addresses throughout a network. If the address format of the different types of network is different, the gateway must handle this as well.

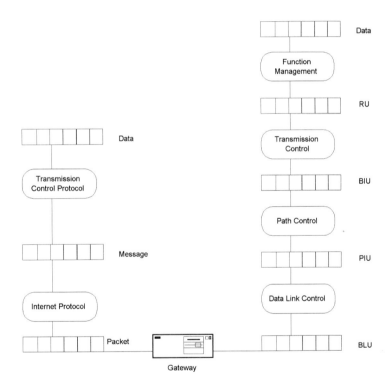

Figure 9-32. Gateways and Datastreams

Gateways and Remote Access

The term "gateway" is also used to describe a router. These gateways are also referred to as remote gateways (See Figure 9-33). For example, Internet routers are called gateways. The term is also used for protocol converters. A modem gateway is another example of a remote access gateway. Modem gateways allow multiple remote asynchronous terminals to connect to a network through modems.

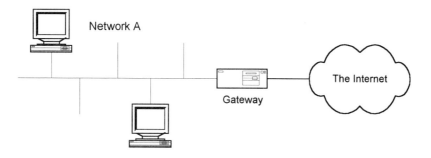

Figure 9-33. Gateways and Remote Access

HUBS

A hub is a network component which centralizes circuit connections. Hubs started out as wiring concentrators but have developed into sophisticated switching centers.

Hub Complexity

One way to categorize hubs is by their level of complexity. Figure 9-34 shows the simplest type of non-switching hub, the Multistation Access Unit (MAU). Network management of an MAU is limited to front-panel LEDs in most cases, and dynamic reconfiguration is out of the question, in other words it is a fixed configuration hub. An MAU supports one type of LAN such as Ethernet or Token Ring.

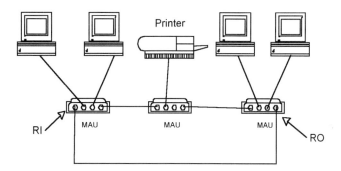

Figure 9-34. Multistation Access Unit

Another type of simple hub is a 10Base-T hub (See Figure 9-35). These hubs are used to con-nect Ethernet and Token Ring networks together in a star configuration. These hubs can be as small as four ports and used for small businesses and home offices, or very large and connected to many other hubs to serve large businesses.

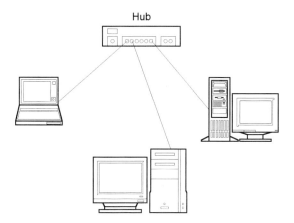

Figure 9-35. 10Base-T Hubs

The next most complex type of hub is another fixed configuration hub (non-switching) which allows remote network management. As with MAUs, these hubs are straightforward and can connect only one type of media.

Adding more sophistication and complexity is the stackable hub, which consists of modules that are stacked one on top of another and are connected by short cables (See Figure 9-36). Network management capabilities exist with these types of devices, but they appear to the network as a single logical device even though physically they are made up of separate modules.

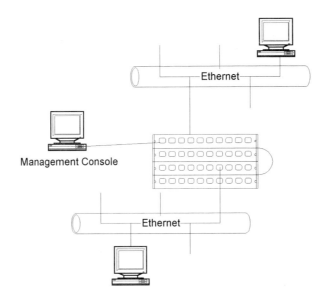

Figure 9-36. Stackable Hubs

Modular multi-LAN hubs are the most complex type of hubs (See Figure 9-37). This type of hub can connect multiple types of LANs and allow users to separate LANs into logical segments. A high speed chassis is used to interconnect individual LANs with network management modules. Internetworking modules also connect to the chassis and support bridge and router capabilities. Hubs of this type are sometimes referred to as "Collapsed Backbones."

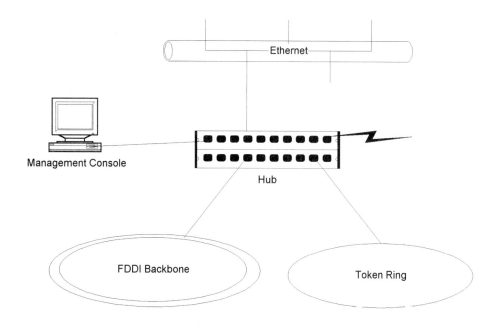

Figure 9-37. Multifunctioning Hub

Port Switching and Segment Switching

Hubs can also be divided into two categories based on switching capabilities, switching hubs, and non-switching hubs (also called fixed configuration hubs). Switching hubs can be further divided into port-switching hubs and segment switching hubs.

Port-switching hubs allow network managers to assign hub ports to network segments through a network management console interface. This has given rise to the term "virtual LAN." This is a simple and easy way to segment a network, but if all ports need to connect to all other ports, this is not a practical solution.

Segment-switching hubs move frames from one port to another, each acting like it's own segment. The source and destination address of the individual frames are monitored and switched as appropriate. This allows for greater connectivity between individual nodes and segments.

SWITCHES

Switches are devices that direct traffic among several networks or network segments. They are also referred to as switching hubs (described briefly in the previous section). The most common type of switch is an Ethernet switch, although Token Ring switches do exist and are gaining popularity. Since Ethernet switches are the most common, they will be used as the subject matter for this section. The function of a switch is to move frames between network segments at a very high rate, thereby increasing the overall bandwidth of the network (See Figure 9-38).

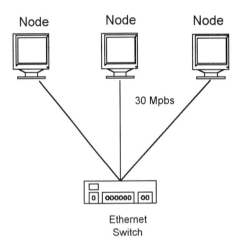

Figure 9-38. Ethernet Switch

An Ethernet switch moves frames between any two attached Ethernet devices on a frame-by-frame basis. The two most common types of Ethernet switches are Store-and-Forward switches and cross-point switches. Store-and-Forward switches check each frame for accuracy and then direct the frame to the destination. Cross-point switches do not error-check the frame, they merely pass the frame to the appropriate destination. Cross-point switches are much faster than Store-and-Forward switches.

Switches can be integrated into existing networks as shown in Figure 9-39. Note the difference in bandwidths available for the devices attached to the hubs versus the devices attached to the switch. This is the fundamental difference between a hub and a switch. A switch is a high performance device capable of switching between segments, each segment capable of 10Mbps for standard Ethernet. The bridge keeps traffic isolated between devices attached to the hub and devices attached to the switch. Users (departments) with greater bandwidth needs would be attached to the switch, while others would use the hub.

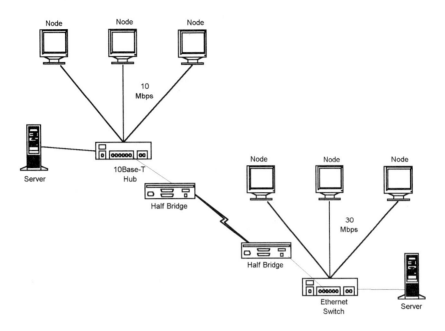

Figure 9-39. Ethernet Switches and Hubs

SIMPLE NETWORK MANAGEMENT PROTOCOL

The 1970s was the decade of the centralized network. In a period dominated by mainframe processing, data communications consisted primarily of a system of terminals connected to a mainframe. Typically low speed, asynchronous transmission was used for the communications link. These mainframe/terminal systems had very little direct interaction with other computer systems or the outside world. Furthermore, support and management of the system was typically supplied by the mainframe manufacturer (IBM) and the communications company (AT&T).

The 1980s brought changes in computer technology that altered the face of traditional data communications. First, microprocessors burst onto the scene. They offered significant price and performance advantages over mainframes and ignited the fuse of the PC explosion. Second, as the number of microcomputers increased, so did the user's desire to share applications and data. This desire led to a dramatic increase in the number of LANs. The increase in LANs caused high-speed wide area transmission facilities, such as T-carrier circuits, to emerge and to provide connectivity between the microcomputer-based LANs.

When the 1980s rolled around, network managers were left with a big problem. Not only was there a connectivity problem between the myriad of products, but there was a bigger problem of how to manage this internetwork of technologies. This led to the development of network management products. These products were developed around network protocols and were aimed at helping network managers deal with the complexity of everyday computing. One such protocol is the Simple Network Management Protocol (SNMP).

To help understand the role of a network protocol, we need to first look at a network management model. A typical network management system is called a manager/agent model and consists of a manager, a managed system, a database of information and the network protocol (See Figure 9-40).

The manager provides the interface between the human network manager and the devices being managed. It also provides the network management process. The management process does such tasks as measuring traffic on a remote LAN segment or recording the transmission speed and physical address of a router's LAN interface. The manager also includes some type of output, usually graphical, to display management data, historical statistics, and so on. A common example of a graphical display would be a map of the internetwork topology showing the locations of the LAN segments. By selecting a particular segment, the operator can display its current operational status.

The managed system consists of the agent process and the managed objects. The agent process performs network management operations such as setting configuration parameters and current operational statistics for a device on a given LAN. The managed objects include workstations, servers, wiring hubs, and communication circuits. Associated with the managed objects are attributes, which may be statically defined (such as the speed of the interface), dynamic (such as entries in a routing table), or require ongoing measurement (such as the number of packets transmitted in a given time period).

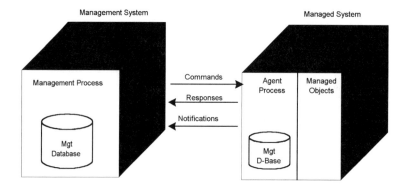

Figure 9-40. SNMP General Model

A database of network management information, called the Management Information Base (MIB), is associated with both the manager and the managed system. Just as a numerical database has the structure for storing and retrieving data, an MIB has a defined organization. This logical organization is called the Structure of Management Information (SMI). The SMI is organized in a tree structure, beginning at the root, with branches that organize the managed objects by logical categories. The MIB represents the managed objects as leaves on the branches.

Most devices on the network do not have the same information in their MIBs for two reasons. First, most devices usually come from different manufacturers who have implemented network management functions in different, but complementary ways. Second, devices perform different internetworking functions, and may not need to store the same information. For example, a workstation may not require routing tables, and would not need to store routing table-related parameters in its MIB. On the other hand, a router's MIB would not contain statistics such as CPU utilization that may be significant to a workstation.

The network management protocol provides a way for the manager, the managed objects, and their agents to communicate. To structure the communications process, the protocol defines specific messages, referred to as commands, responses, and notifications. The manager uses these messages to request specific management information, and the agent uses them to respond. The building blocks of the messages are called Protocol Data Units (PDU). For example, a manager sends a request PDU to retrieve information, and the agent responds with a response PDU.

SNMP is based on the manager/agent model. Its primary purpose is to allow the manager and the agents to communicate. This protocol provides the structure for commands from the manager, notifies the manager of significant events from the agent, and responds to either the manager or agent. The original version of SNMP was derived from the Simple Gateway Monitoring Protocol (SGMP) and was published in 1988. At that time the industry agreed that SNMP would be an interim solution until OSI-based network management using CMIS/CMIP became more mature. Since then, however, SNMP has become more popular than the OSI solution and has been more widely adopted than originally anticipated.

Overall, SNMP is designed to be simple. It does this three ways. First, by reducing the development cost of the agent software, SNMP has decreased the burden on vendors who wish to support the protocol. This increases the acceptance of the SNMP. Second, SNMP is extensible because it allows vendors to add their own network management functions. Third, it separates the management architecture from the architecture of network devices such as workstations and routers. This further widens the multivendor acceptance and support for this protocol.

SNMP is also referred to as simple because the agent requires minimal software. Most of the processing power and data storage resides on the management system, while a complimentary subset of those functions resides in the managed system. To achieve its goal of being simple, SNMP includes a limited set of management commands and responses (See Figure 9-41).

The management system issues Get, GetNext, and Set messages to retrieve single or multiple object variables or to establish the value of a single variable. The managed system sends a response message to complete the Get, GetNext, or Set. The managed system sends an event notification, called a trap, to the management system to identify the occurrence of conditions such as a threshold, that exceed a predetermined value.

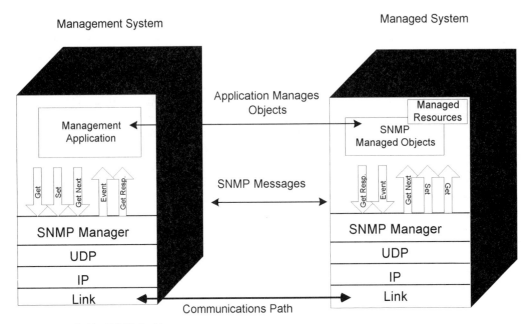

Figure 9-41. SNMP Architecture

Because most management information does not demand the reliable delivery that connection-oriented systems provide, the communication channel between the SNMP manager and agent is connectionless. In other words, no prearranged communication path is established before the transmission of data. Neither the manager nor the agent relies on the other for its operation. Consequently, a manager may continue to function even if a remote agent fails. When the agent resumes functioning, it can send a trap to the manager, notifying it of its change in operational status. Even though SNMP makes no guarantees about the reliable delivery of the data, in reality most messages get through, and those that do not can be retransmitted.

The most common implementation of SNMP is in the TCP/IP world. It relies on the User Datagram Protocol (UDP) and the Internet Protocol (IP) for proper operation. SNMP also requires Data Link Layer protocols, such as Ethernet or Token Ring, to set up the communication channel from the management to the managed system.

SNMP has a very straightforward architecture. Figure 9-42 compares the SNMP architecture to the OSI Model. The manager/agent applications reside at the Application Layer. Below the Application Layer is the Presentation Layer. The ASN.1 encoding resides at this layer and ensures uniform structure of management information and proper syntax for the SNMP messages. SNMP's connectionless communication mechanism removes some of the need for a Session Layer and reduces the responsibilities of the lower layers. For most implementations, the UDP performs the Transport Layer functions and the IP provides the Network Layer functions, and LANs such as Ethernet and Token Ring provide the Data Link and Physical Layer functions.

The use of SNMP agents within internetworking devices has increased dramatically in the last few years. There are five general categories of devices in which you will find agents: wiring hubs, network servers and their associated operating systems, network interface cards and the associated hosts, internetworking devices such as bridges and hubs, and test equipment such as network monitors and analyzers. Other devices such as uninterruptible power supplies have also become SNMP compatible.

What is the future for SNMP? Since its inception in the late 1980s, SNMP has enjoyed an eagerly growing user base. With support from major vendors like IBM, DEC, and HP, SNMP is bound to continue growing. SNMP creators have recently bolstered its arsenal by releasing SNMP version 2. This release addressed the problems found in version 1 and expanded the capabilities of the protocol.

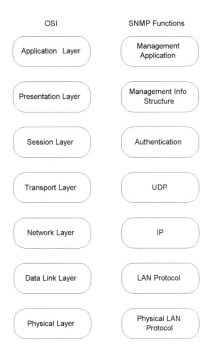

Figure 9-42. SNMP and the OSI Model.

SELECTING INTERNETWORKING COMPONENTS

This section analyzes different networking components and the advantages and disadvantages of each. When to use a repeater, bridge, or router is often a difficult question to answer. We start with the simplest and most straightforward component, the repeater, and proceed through the most complex, the multiprotocol gateway. Once we have looked at the advantages and disadvantages, we consider some key questions to ask before purchasing each product.

Repeater

Advantages

1. Simple to operate: repeaters do not provide very much functionality and therefore are simple devices compared to other internetworking devices
2. Low latency: because repeaters are hardware intensive devices, they are able to operate at very fast speeds. Repeaters simply copy bits and do not contain the intelligence to interpret addresses or make routing decisions
3. Inexpensive: repeaters are typically the least expensive internetworking device type
4. Extends the distance of a LAN segment beyond cable length restrictions
5. Extends the distance some types of workstations may be from LAN segments or backbones

Disadvantages

1. Limited functionality
2. Sometimes limited management capabilities

Repeater Considerations

1. Latency
2. Cable type supported

Bridge

Advantages

1. Provide traffic isolation which increases LAN segment performance
2. Operate independently of upper layer protocols
3. Regenerate signal levels

4. Provide WAN connectivity (Remote Bridges)
5. Usually less expensive than routers
6. Low latency compared to routers

Disadvantages

1. Cannot be used to route traffic across large networks or multiple networks
2. Will not convert protocols above the data link layer

Bridge Considerations

1. What you get in the base configuration
2. Hardware upgrade costs
3. Software upgrade costs
4. LAN Architectures supported
5. Remote access support for WAN bridges
6. Management capabilities
7. Performance (frames per second or transmission speeds supported)

Router

Advantages

1. Can route packets from source to destination through complex networks
2. Can be configured to allow for multiple paths
3. Allows for network data flow control
4. Allows access to large networks such as the Internet

Disadvantages

1. High latency compared to a bridge
2. Cost compared to a repeater or bridge

Router Considerations

1. What you get in the base configuration
2. Hardware upgrade costs

3. Software upgrade costs
4. Whether the router will handle multiple protocols or a single protocol
5. Network connectivity support
6. Performance characteristics (Packets Per Second)
7. Management capabilities (i.e., SNMP support)
8. LAN Support (Ethernet, Token Ring, FDDI, etc.)
9. Latency

Hub

Advantages

1. Relatively easy to implement wiring changes
2. Network expansion is made easy
3. Most hubs support multiple LAN and WAN protocols
4. Centralized management
5. Provide repeater capabilities

Disadvantages

1. Cost
2. Possibly a single point of failure

Hub Considerations

1. Redundancy features
2. LAN Topology support
3. Port switching capabilities
4. Segment switching capabilities
5. Migration capabilities

Gateway

Advantages

1. Converts from one architecture to another

Disadvantages

1. Very high latency

Gateway Considerations

1. What you get in the base configuration
2. Hardware upgrade costs
3. Software upgrade costs
4. System Architectures supported
5. Management capabilities
6. Performance

SUMMARY

Repeaters operate at the Physical Layer and reproduce bit for bit everything they receive. They can extend the length of a LAN and connect different media types.

Bridges operate at the Data Link Layer and are transparent to the Network Layer. They can isolate one part of a network from another to reduce LAN traffic. They can create parallel routes for data and can extend a network over a wide area. They can interconnect LANs and can be used to form network backbones. There are two types of bridges, Source Route Bridges and Spanning Tree Bridges.

Routers operate at the Network Layer of the communications architecture and are transparent to the upper layers. Routers interconnect networks that have the same communications architectures. A multiprotocol router handles multiple communications architectures but does not convert protocols. Two half-routers or WAN routers connect two networks transparently across a point-to-point link. Routers use routing tables to route packets. TCP/IP routers use the EGP protocol to communicate routing information to core gateways on the Internet backbone.

Hubs are network components which centralize circuit connections. Hubs started out as wiring concentrators but have developed into sophisticated switching centers.

Switches are high-performance hubs which increase the bandwidth of networks. There are two primary types of switches, Store-and-Forward or Cross-Point.

Gateways or protocol converters lie at the boundary of systems which do not use the same communications architecture. They operate at the OSI Network layer and above. A gateway accepts a data stream which has data encapsulated by several layers of one communications architecture and must construct an encapsulated message with all of the headers required for the other architecture.

ADDITIONAL INFORMATION ON THE CD-ROM

- Additional Information on Subjects found in this chapter
- SNMP Traces
- Additional Network Configurations
- Additional Component Specifications

10

THE INTERNET

CHAPTER OBJECTIVES

The purpose of this section is to introduce you to the Internet and tools available for the Internet. The topics covered in this section are:

- History of the Internet
- Ways to Access the Internet
- Navigation Tools Used on the Internet
- Internet Issues
- The Future of the Internet
- The World Wide Web
- The Mosaic

HISTORY OF THE INTERNET

The Internet was started in 1969 by the Department of Defense to maintain military communications in the event of major disruption (nuclear war) in the telephone service. It linked universities and military contractors by leased telephone lines. Each computer site followed agreed-upon guidelines created by a standards committee. The government provided funding for the Internet for several years and allowed only nonprofit, educational, and government use.

In 1991, the National Science Foundation (NSF), who pays for approximately 10 percent of the leased lines, loosened the guidelines for Internet usage. They started allowing many commercial uses, such as announcements of new products or services, for use in research or instruction, but not advertising. This change flooded the Internet and has caused enormous growth from the inundation of commercial messages on non-NSF lines. The Internet continues to grow at an enormous rate and new users continue to be connected every month.

The Internet hosts the broadest e-mail system in existence. A user can carry on interactive conversations with other users and find out almost anything about anything on the Internet. There is a mass of information on the Internet such as:

- Libraries of electronic books
- Public Domain Software
- Technical papers
- *The Wall Street Journal*
- Hourly world news reports
- Songs, recipes, poetry, etc.

The Internet is considered by some to be the "Information Superhighway." It is a global network of networks. The Internet is actually the physical cables and computers of the global network. Proprietary Wide Area Networks (WANs) are much more expensive than the Internet. The Internet provides many of the capabilities of WANs at a fraction of the cost and has become a primary WAN connectivity option for connecting corporate networks.

Ways to Access the Internet

Getting to information on the Internet requires gaining access to the Internet backbone via gateways which are connected to the backbone. Once you have access to the backbone you can get to all other services which are connected to the Internet. Getting access to the backbone requires using the services of an access provider. There are several ways to access the Internet backbone. Listed below are the most common methods:

Dedicated Circuit

A full-time network connection setup by the local phone companies can be used for access. Speeds can vary depending on needs. Common speeds are 56Kbps and T1 to the Internet Access Provider.

SLIP and PPP

SLIP (Serial Line Interface Protocol) uses standard modems and switched phone lines. Software is required on your PC to run the SLIP protocol. It has the same functionality as the dedicated

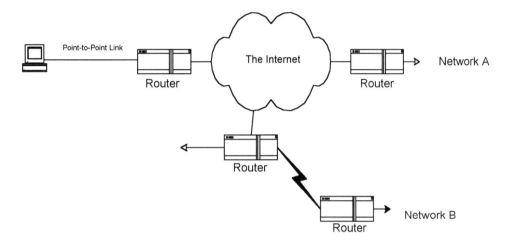

Figure 10-1. Point-to-Point Internet Connectivity

circuit but at slower speeds such as 9.6Kbps and 19.2Kbps (See Figure 10-1). The Point-to-Point Protocol (PPP) is another example of a commonly used protocol for this type of connectivity. Both SLIP and PPP provide connectivity to an Internet Access Provider who connects you to the Internet Backbone.

UUCP (UNIX to UNIX Copy)

UUCP uses standard modems and switched phone lines. The primary limitation is that it does not provide a means for connecting to interactive hosts on the Internet. You can still send mail and transfer files.

Interactive Connection

When using dialup with modem access, you must connect to another Internet host and use the facilities of that host.

Navigation tools

WAIS, Archie, TELNET, Gopher, FTP, mail, and World Wide Web are some of the Internet tools used to access information on the Internet. NetScape and Mosaic are graphical Windows interface tools that make it easier to connect to Gopher, the Web, etc., and navigate the Internet. Following is a brief description of some of the navigation tools, but the focus in this section will be on the Web.

- Archie is a network service that searches FTP sites for files. It is a program that lists the contents of the major FTP sites.

- FTP (File Transfer Protocol) is a common method of transferring files across networks.

- Gopher is a versatile, menu-driven information service. It is similar to the Web, but not as powerful.

- USENET is the global news-reading network.

- Veronica is a network service that allows users to search Gopher systems for documents.

- WAIS (Wide Area Information Service) is a service which allows users to intelligently search for information among databases throughout the network. It is a full-text database system. Database systems allow retrieval of documents by specifying any of the words occurring in the documents.

Internet Issues

The Internet does not have a starting point or top. There is no police department monitoring the highways. This results in somewhat chaotic traffic flow. Therefore, some of the more salient issues are:

- Some of the content of the information lacks a business or educational focus.

- Fear of unauthorized access to Internet files makes security an issue. This has been addressed somewhat with firewalls (software, hardware, and physical breaks between Internet gateways and LANs), data encryption, and passwords.

- There are concerns with the usage of hypermedia and e-mail with voice messages flooding the Internet. It will not be able to deal with it and will require increased bandwidth.

- The Internet is not a smooth superhighway. It is not easy to navigate, has bumpy roads, and does not always have road maps or the fastest cars.

- Productivity may decrease due to employees non-work related "surfing" of the net.

The Future of the Internet

The current United States administration has committed itself to creating an information super-highway for the twenty-first century. The administration wants everyone to be able to afford access to this information. It is unknown at this time what these highways will cost.

The World Wide Web

Internet resources are managed and downloaded by World Wide Web (WWW) servers. The Web is a distributed hyper-text based information system developed at CERN by Tim Burners-Lee in 1989.

The Web refers to the information on the Internet. It allows you to obtain information. It is based on the concept of hypertext and hypermedia, in which servers exist solely to link users to documents and multimedia files. Hypertext is computer information containing text that can be linked with selected phrases. The links point to other documents or files. It is basically the same as regular text, but it contains connections within the text to other documents. The links in the text are called hyperlinks. Hypermedia takes hypertext another step to include images, sounds, and video with links that can be selected and viewed.

The Web allows an easy interchange of hypermedia between networked environments. Some of these environments or clients can be terminal browsers, PCs running Windows, Macintoshes, or X-Windows on workstations. The Web is free and in the public domain.

The Web works under the client-server model. A Web server is a program running on a computer whose only purpose is to serve documents to other computers when asked to. A Web client (also called a browser) is a program that interfaces with a user and requests documents from a server as the user asks for them. Web clients are available on text-only terminals, UNIX, VMS, Macs, PCs, etc. Web servers are available on PCs, UNIX, Perl, Macintosh, VM, and VMS.

The Web understands the following interfaces through hypertext and some hypermedia for anything served through Gopher, WAIS, FTP, Archie, Veronica, USENET, TELNET and other protocols. It understands ASCII, GIF, postscript, and other data formats. The Web uses a new protocol, (HTTP) and data format, Hypertext Markup Language (HTML).

Information can be downloaded and uploaded to the Web. An HTTP server, Gopher server, FTP server, or WAIS database server is needed to load information onto the Web. To use hyper-text with reasonable speed, an HTTP server should be used.

Hypertext Markup Language (HTML) is the language the Web uses to create and recognize hypermedia documents. HTML is known for its ease of use. HTML documents are 7-bit ASCII files with formatting codes that contain information about layout (text styles, document title, paragraphs, lists) and hyperlinks.

HTML uses Uniform Resource Locators (URLs) to represent hypermedia links and links to network services within documents. Documents on the Web are referred to as URLs. A URL looks like http://www.vuw.ac.nz/campus/home.html. It consists of three parts — the method of retrieving the document (HTTP), the address of the computer where the data or service is located (www.vv.vuw.ac.nz), and a pathname of the port to connect to or the name of the file or text to search for in a database (/campus/home.html). When a hypertext link is selected in an HTML document, the user is actually sending a request to open a URL. The URL format is nearly an Internet standard.

In 1992, several client/server applications were available for browsing the information net including the World Wide Web. The National Center for Supercomputing Applications (NCSA), a consortium-based research facility, became interested in CERN's project. NCSA's missions are to provide easily available, noncommercial software to the research community, and investigate new research technologies for commercial interests. Creating an interface to the Web with enhanced client capabilities appeared to be appropriate.

Mosaic

NCSA initiated Mosaic, a Web-inspired client application used to access the vast information on the information net. The design contained many features including a GUI interface, traditional query protocols, multimedia capabilities, and ease-of-use capabilities. It lets you escape the low-levels of query protocols to easily search the information net.

Features of Mosaic

Mosaic includes basic features that are readily appreciated. The first is that Mosaic is a Public Domain application (free) and the user can download the application using FTP. When ready, the user discovers mouse point-and-click capabilities and a screen that is familiar to Windows users. The pull-down menu includes screen customization, edit options, and print capabilities. The filing system allows for retrieval and storage of files locally or on the Internet. Other areas are customized for the exploration of the information net.

The "Navigator" assists in charting the unknown territory. As the user points and clicks their way through highlighted words and phrases of Hypertext, the user's list creates a map, charting their footsteps. When something of interest is discovered, a book mark is set, designating the document for future reference. The Uniform Resource Locator (URL) information is placed on the user's "hot list", creating a personal reference guide.

Reading an entire article can be extremely time consuming when searching for specific details. Mosaic includes word search capabilities to assist the user in scanning documents quickly.

One small but significant feature is Mosaic's trademark, the spinning globe that appears in the upper right-hand corner of the screen (See Figure 10-2). The globe spins as it is browsing for information so the user gets the feeling that something is happening. But what if nothing happens and the system seems to have quit? Resetting or powering down are always options; but with a click on the globe, Mosaic drops the search and the user can continue from the point prior to the problem.

Creating Documents for the Information Net

Mosaic documents are created in the Web's HTML ASCII format. HTML can translate documents transferred from (and to) text editors including Frame Maker, Interleaf, WordPerfect, MSWord, Troff/Nroff, etc. The document structure includes the option to include comments within the text. The title becomes the reference for indexing in other areas. The creator can easily examine and read the document thanks to the lack of cryptic notations needed to create the document.

For reference, anchors are used to hyperlink documents to additional information. Specific sections of text in a referenced document can be reached through additional URL syntax. Multiple types of lists can be created and referenced. Libraries are available for use to include sound, images, and even motion pictures. The document is created with hypertext to locate the media to use. With the anchor set, the client clicks and experiences an information network document.

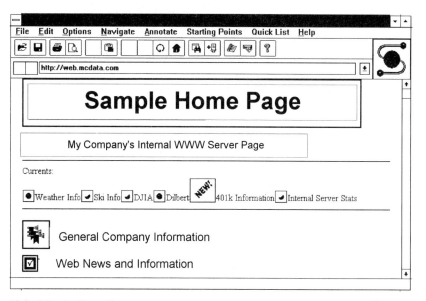

Figure 10-2. Mosaic Home Page

URL

The URL was designed to interpret the various schemes employed on servers in the information net. The URL specifications are hazy and very generalized to adapt to the appropriate scheme.

URLs create an anchor that can reference another document, sound file, news server, image, or movie. The text is highlighted and the user just points and clicks the information the URL is pointing to as presented.

The URL can present different information depending on what is requested. Mosaic will respond depending on the level that the information is being requested. When a file is requested, Mosaic will try to display, or play the sound or video, that has been requested. If the URL requests information at the directory level, it will display the directory. At the server level, a directory of what is located on the server is displayed. At this point, if the user did not know where the information was located, a hierarchical search could be conducted. Mosaic does not have the capability to display the hierarchical structure of a server or directory.

Costs

Even though Mosaic is a Public Domain Application (free), there are always costs. Mosaic is available for use with most operating system.

The network uses the TCP/IP protocol and network applications rely on SLIP and PPP for setup. Many business have established TCP/IP networks; however, the small business and individual Internet client might need to investigate these protocols.

Each protocol furnishes a login and packet encapsulation scheme for transmission over serial lines. SLIP is provided in many UNIX networking options and is available via FTP for various operating systems and hardware. SLIP uses its own framing and packet formats implementing TCP/IP over the serial lines. PPP provides greater throughput resulting from compression and encapsulation schemes. Both low-speed serial lines and high-speed wide area links can be used, allowing network protocols to share the same line, including TCP/IP, OSI, IPX, SPX, AppleTalk, and others. The login links can be a dedicated circuit, voice grade phone line, or possibly a satellite link (which may become common in the future).

A modem with a voice grade phone line is all that is required for the interfaces. However, carriers which allow partitioning of T1 links into fractional speeds of 56K, 128K, 256K, and so on should be evaluated. The leased higher-speed T1 services could ultimately be more economical due to distance from the vendor's network access point (the further away you are the higher the price) and the installation of fiber-optic cable. Different quality levels of high-speed network services are also available.

NCSA provides online documentation and help facilities that are easily available while using Mosaic, if you are not behind a firewall. The firewall does not allow access outside of the users area. The consequences involve searching via some other protocol if it has external permission. Some documentation is included with the file transfer through text files with ideal names such as README.

What Do You Need Mosaic For?

Information pertinent to a business or individual must be located and retrieved in order to be exploited. The challenge is to find a tool that automates the process of pinpointing the needed material. Mosaic provides simplified tools for locating material on the net and providing information to the net.

The ability to create information for the net provides many opportunities to the client. A company can establish its own information network for internal access or restricted external access. Information is available by using the home page concept instead of advising someone of a path name. Information is current and easily accessible without passing files. The annotate features give an additional collaborative tool. Distances between vendors, plants, and departments shrink with increased usage.

Establishing an open server on the information network connects your business to the rest of the world. Some of the common usages for Web Servers are:

Product announcements: Information is spread globally for pre-sales and free information. With the multimedia of Mosaic and HTTP, demos can be presented to the shopper through a variety of advertisements.

Post Sales Support / Updates: Support for products that have been sold can be accomplished through the information net. Instructions on installation, implementation, or enhancements are easily communicated. Customers are presented with a closer link to their product information and the annotate facility can be a great asset for this type of support.

Online Information Delivery and Help Systems: When will it be delivered? Instead of using the telephone for this question, customers and vendors will be able to check on the information net for your company. Problems can arise with shipments and notification of problems is easily posted or sent through the e-mail system.

SUMMARY

The Internet is the fastest growing network in the world. It uses the TCP/IP protocols as the basis for communications. There are many tools used to navigate the Internet and retrieve information. One of the most common tools is Mosaic, a graphical user interface tool which makes navigation simple. The Internet has become an economical way to create Wide Area Networks (WANs) between businesses.

ADDITIONAL INFORMATION ON THE CD-ROM

- Internet Request for Comments (RFCs) such as:
    - Internet Protocol RFCs
    - Frequently Asked Questions
    - Internet Protocols
- Additional information on:
    - Archie
    - Gopher
    - WAIS
    - Finding People on the Internet
    - Mosaic
- SLIP Protocol Description
- PPP Protocol Description
- Internet Specific Definitions
- TCP/IP Protocol Information

11

TELECOMMUNICATIONS

INTRODUCTION TO TELECOMMUNICATIONS

Up to this point we have treated the communications channel as a "wire," assuming that the OSI Layer processes simply transmit bits across it. The exception was in Chapter 4, where we discussed the technology used for the communications channels in LANs. That made sense, because LAN technology spans Layers 1 and 2, and, to a large extent, the basic concepts are intertwined with the technology used for the channels.

We were able to ignore other telecommunications technology, that is, the technology of long distance channels, because the communications architectures effectively isolate them from the higher layers. The channels are not really just wires, of course. They are, in fact, complex networks with their own communications architectures and specialized technology. Anyone selling or supporting communications products, even those that fit in at Layers 2 and 3, will often be exposed to both the technology and the economic issues of the public switched networks and similar private networks.

This chapter will introduce you to components found in telephony, the technology of switched voice communications. This is necessary background information for the following discussion of how the communications channels of the public switched networks are used for data communications, and voice and data communications have been integrated. We will tell you about interesting new technology that further integrates them. We will end by introducing you to the major providers of telecommunications services and discussing the tariffs they charge for their services.

Throughout this chapter, we will use the term subscriber to denote those who use the telephone networks, rather than user or node, as we have in previous chapters. By subscriber, we mean the organization or enterprise at a given site, but at times it will be clear from the context that we mean either an individual user or all of the sites of an enterprise. Considered from the broadest perspective, there are four parts to the evolution of today's telecommunications networks:

1. The development of analog voice networks during a 100-year period that began in the mid-nineteenth century

2. The adaptation of analog voice networks to the transmission of digital data between computers and terminals from the late 1950s onward

3. The conversion of the analog voice networks to digital voice transmission technology beginning in the 1960s and ending, in the U.S., in 1980

4. The adaptation of digital voice networks to data communications, beginning in the early 1980s with the use of T1 for data and continuing in the 1990s with technologies such as Integrated Services Digital Networks (ISDN) and Asynchronous Transfer Mode (ATM)

CHAPTER OBJECTIVES

At the completion of this chapter, you should be able to:

- Indicate the usage and definitions of common terms used in telephony

- Indicate frequency ranges for human speech and for analog voice channel frequencies

- Indicate the correct nomenclature for the various levels of the public switched network switching hierarchy

- Identify the line signal standards used throughout the world and indicate the bit throughput for those standards used for data communications

- Specify the major components of a data communications link and their interface points

- Identify the sublayers of the physical layer and indicate common protocols used by those sublayers

- Identify the major characteristics of modems and modem protocols

- Identify the important elements of digital telecommunications systems

- Identify the X.25 protocols for Layers 1, 2, and 3, and indicate the service each provides

- Specify the major ISDN channels, protocols, services, and subscriber-site components

- Identify emerging telecommunications architectures and technologies and specify how they fit into the telecommunications picture

- Differentiate between cell relay and frame relay and identify major characteristics of each

- Determine the major characteristics and capabilities of an IEEE 802.6/DQDB MAN

- Indicate the interfaces and protocols for SMDS

- Differentiate between various phone company entities and give their major interface points

A Bit of History

Existing analog telephone networks in the United States and indeed the world represent many remarkable engineering achievements.

In 1835, after three years of experiments, Samuel Finley Breese Morse completed his first model of an electromagnetic telegraph apparatus. This model consisted of a sender to transmit signals by the opening and closing of an electric circuit, an electromagnetic receiver to record the signals, and a coding algorithm which translated the signals into letters and numbers. He later invented a system of electromagnetic relays to transmit the messages over longer distances.

Joseph Henry provided some technical assistance and he had financial backing from one of that eras most prominent bankers. In 1843, the United States Congress voted a $30,000 appropriation to build an experimental line from Washington D.C. to Baltimore, Maryland, seven years after he had applied for a patent. A year later, he sent a message on that experimental line. He formed his own company to market his invention when the U.S. Government failed to buy the rights to the invention. He was involved in a series of costly lawsuits which he eventually won. His invention came to provide a substantial communications infrastructure in the world. In effect, this system represents the first electrical digital transmission system.

In 1874, Alexander Graham Bell determined that sound could be transmitted over a wire, by electricity, through the variation of the intensity of the current, corresponding to the variations in air density produced by sound. He filed for a patent on February 14, 1876 for the invention of the telephone. On March 10, 1876, the first sentence ever transmitted by electricity over wires was sent by him.

The telephone was of little use without some means of easily changing connections. In 1878, in New Haven, Connecticut, the first switching office was established. This was a precursor to what are now referred to as Central Offices.

To give you an idea of how far we have now come, the following statistics were taken from the Web Pages of ATT:

- The ATT Network handles more than 185 million voice, data, and image calls each business day.
- It reaches more than 270 countries and territories.
- It provides direct dial access to more than 200 countries and territories.
- It offers the fastest call setup time in the industry.
- It is virtually 100 percent digital.
- It includes more than 2.75 billion circuit miles of transmission facilities.
- It transmits more than 95 percent of all traffic over fiber optic circuits.
- It ensures network reliability by using the patented FASTAR(SM) system to automatically route calls around fiber-optic cable failures.

Today's Networks

Telecommunication networks have evolved to find applications in virtually all areas of modern society: banking and finance (remote teller banking systems, electronic funds transfer networks, stock brokerage networks, and credit verification systems), reservation systems (the airline and travel industry, hotels and motels, sporting events, concerts, etc.), grocery and retail store check-out systems and home shopping, and offices and factories (automation, time-shared computing in corporate networks, and electronic mail systems). Personal computers with modems now access a wide variety of networks and provide a major impetus for network growth.

A few examples illustrate the scale of today's networks. General Motors currently operates a network that links together more than 500,000 computing devices and telephones and connects 18,000 locations worldwide. American Airline's SABRE reservations network links more than 60,000 video terminals all over the globe to six large-scale computers (processing about 1800 messages per second, or 60 million messages per day). Sometimes the SABRE network posts larger annual revenues than the airline itself.

Today, communications systems that send voice, data, and video signals span the earth. Information is bounced off of satellites and directed through cables that cross entire oceans. Computer-based systems operate in the dead of night, receiving messages from places where it is high noon. The organization that turns a blind eye to how communications and computer technologies together are transforming the world will be outperformed and outmaneuvered by its competitors—competitors that can now be found in any country on earth.

A Look to the Future

A futuristic view of the potential in the telecommunications field would include the following facilities or services:

At Home:

- Virtual Theaters—video, movies on demand at home

- Virtual Stores/Malls—interactive shopping, banking at home

- Virtual Stadiums—sports events on demand at home

- Virtual Newspapers—multimedia press

At Work:

- Virtual Workgroups—interactive, concurrent, collaborative design

- Virtual Conference Rooms—teleconferences at the desktop

- Virtual Libraries— multimedia libraries

- Virtual Schools—interactive teleseminars

- Virtual Hospitals—remote diagnostics in real-time

Realization of the above has already occurred in many cases. Others have been limited by legal, regulatory and economic factors, and by basic human nature rather than by the availability of technology.

TELECOMMUNICATIONS SERVICES AND CONCEPTS

When AT&T was divested of its local phone companies on January 1, 1984, two distinct types of telephone companies were legally defined (Refer to Figure 11-1):

1. Local Exchange Carriers (LECs). These consist of the twenty-three Bell Operating Companies (BOCs) that were created by the divestiture, the former independent telephone companies such as GTE and Contel, and about 1500 small-town telephone companies.

2. Interexchange Carriers (IECs or sometimes IXCs), more often simply called long distance carriers. These are the former AT&T Long Lines organization and other carriers, such as MCI
and Sprint.

The U.S. was partitioned into Local Access and Transport Areas (LATAs). LATA boundaries
conform more or less to the standard metropolitan statistical areas defined by the U.S.
Department of Commerce. Originally, they conformed to the boundaries of the areas served by
the BOCs.

LECs are prohibited from carrying inter-LATA calls, and IECs are prohibited from carrying
intra-LATA calls. Even when an LEC covers three LATAs, as in the case of Mountain Bell in
Colorado (the 303, 970 and 719 area codes represent three LATAs), it must route the call through
an IEC. An IEC is represented in a LATA by a Point of Presence (POP). A subscriber can connect to the IEC's facilities only at a POP. Typically, this connection is made by a line provided
by the LEC from the subscriber's premises to the POP (See Figure 11-2). Alternately, the subscriber can bypass the LEC and connect directly to the POP. For example, the subscriber could
install a microwave link to the POP (B).

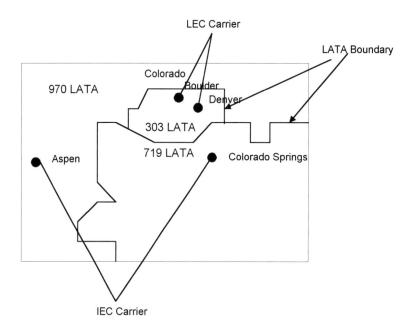

Figure 11-1. Local Exchange Carriers

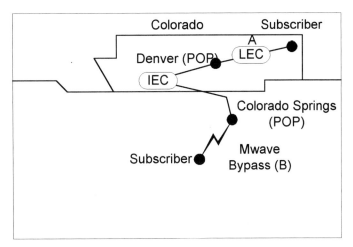

Figure 11-2. Interexchange Carriers

The providers of inter-LATA services fall into two classes:

- Long Distance Carriers. Companies that provide long distance private lines or virtual private networks (switched lines used in the same manner as dedicated private lines).

- Packet Carriers. Companies that provide packet switching services. These companies often use the facilities of the long distance carriers to construct their networks.

Long Distance Carriers

Long distance carriers compete with offerings in these specific categories:

- Voice Grade Service provides leased private analog lines.

- Digital Data Service (DDS) provides leased digital lines used for data only at speeds ranging from 2400 bps to 19.2 Kbps or, in some cases, 64 Kbps.

- DS0 Service provides 64 Kbps digital lines, each providing one DS0 channel. Some carriers provide a clear channel capability, meaning that the carrier does not require any portion of the channel for control signaling. The channel is "clear," and the entire 64Kbps bandwidth can be used by the subscriber. Otherwise, only 56 Kbps of the 64 Kbps can be used.

- Fractional T1 lines will be discussed later in this chapter. Some vendors allow the individual FT1 components to be accessed as if they were private voice grade lines or DDS lines. Others provide only T1 access.

- T1 Service provides lines that support voice, data, or video at DS1 speed. Service may be channelized or unchannelized. Channelized means that it is divided into channels by the carrier, and the carrier takes care of multiplexing multiple DS0 channels onto it. If unchannelized, the subscriber is responsible for multiplexing.

- T3 Service is like T1 service, but at DS3 speed. Generally unchannelized, but sometimes offered as a channelized service.

- International Private Lines of the various types are provided by some carriers to international locations through certain gateway cities. The subscriber must lease two lines, one to the gateway and one from the gateway to the international location.

- Switched Data Services are often referred to as virtual private networks (VPNs). They allow the subscriber to make use of the circuit switching capabilities of the carrier's long distance facilities to control and monitor their private network. Typically, the carrier provides the subscriber with terminals which interface with the carriers network control facility. The subscriber can monitor and reconfigure the VPN within certain limits established by the carrier.

Packet Carriers

Services provided by packet carriers originated in the 1960s, but came into their own in the 1970s with the definition of X.25. Before X.25, each vendor used a proprietary protocol. Users were unwilling to invest in the effort required to utilize proprietary networks.

These carriers have two basic products to offer:

1. Economical Network Access. Some subscribers have relatively low volumes of message traffic or message volumes that fluctuate widely. By allowing subscribers to share transmission facilities and to pay for usage based on message traffic and connect time, the carriers provide an economical alternative to acquiring leased or circuit-switched lines from the long distance carriers.

2. Value Added Services. All of these vendors have enhanced their networks to provide many services in addition to just X.25 packet transmission. For example, nearly all offer electronic mail, not only to any user on their network, but also to users on other networks.

Services provided by these carriers are differentiated by these factors:

1. Type of Access. Most offer dedicated access (essentially a leased line from the subscriber's premises to the nearest point of access to the network). Many also offer dial access, which can be public or private (only one subscriber uses the access lines).

2. Access Speed. Most offer 9600 bps dial lines and 64 Kbps dedicated lines. Packet movement within the network is, of course, at 64 Kbps regardless.

3. Protocols. Virtually all support X.25. A wide variety of other protocols are supported, including SNA 3270 and X.400. These protocols are often linked to the value-added services offered. Recently, several of the vendors have used X.400 to interconnect their electronic mail services.

4. Dial Availability. These services are only economical for dial users if there is a local number to dial. The number of such access points varies a great deal from provider to provider. Path coverage in the continental U.S. ranges from a few hundred cities to over one thousand. Some vendors also provide international access.

Tariffs

A tariff is a document that describes the services offered by a carrier and defines the price for each service. If the services are regulated, then the tariff is filed with the regulating agency in order to place it in force, and of course the agency must review and approve it before it becomes effective.

For IECs and LECs, tariffs for a given communications link are generally structured as follows:

- A base charge for the availability of the link

- A per-mile charge, which typically decreases in steps as the distance increases

- A charge for any optional services or capabilities, such as switched backup or network management capabilities

Tariffs for packet switching are usage-oriented. Carriers typically charge a flat rate for accessing the network plus a charge for each packet sent. That charge will often decrease with volume and increase with distance.

An important factor in the effective cost of establishing a link is the cost of connecting to the POP. The length of the lines between the subscriber's premises and the local POPs at either end might be a small fraction of the overall length of the link, but the tariffs for those lines might equal or exceed the cost of the long distance carrier's lines. There are several reasons for this, including the fact that the local loops are often truly dedicated cables, while the long distance link will typically be multiplexed with many others. When local loop charges are excessive, the subscriber has the option in some cases of bypassing the LEC, as mentioned earlier, with microwave, satellites, or even private fiber optic cable links.

Determining the least costly solution to communications needs is a complex process for the subscriber. In addition to the complexity of the tariff structures, charges differ widely between carriers.

Further complicating tariff issues is the bundling of services. For example, in 1985, AT&T introduced AT&T Tariff FCC #12-Custom-Designed Integrated Services (CDIS), which included leased lines and switched services for voice, data, and video. Commercial subscribers to CDIS are typically large corporations, and the multiyear value of CDIS contracts ranges from under ten million dollars to nearly one-half billion dollars. CDIS has the effect of locking in large customers and has become the subject of lawsuits by AT&Ts competitors and by a communications users group.

Telecommunications Reform Act of 1996

The Telecommunications Reform Act of 1996 opened up competition between the RBOCs and the long distance carriers. This act established guidelines which allow the RBOCs to enter the long distance service market and the long distance carriers to enter the local loop. There are two key policies established by this bill:

1. Wholesaling. Carriers must now allow others to resell existing services which opens up the door to more value-added service providers and discounted service providers. An example of a value-added service would be bundling local, long distance and internet access services.

2. Interconnect. This policy prohibits RBOCs and other companies from charging other carriers unreasonable rates for call terminations. It is thought that this type of "rate protection" will allow competition (such as from cable companies) and ultimately lower prices.

Analog Voice Networks

With the advent of direct distance dialing (area codes) in the 1960s, it became possible to obtain a high-quality connection to any of over 100 million telephones in the U.S. by simply dialing the address (phone number) of that device. This was the culmination of 100 years of development of electromechanical devices for the routing of calls and of analog devices for the transmission of voices.

Voices were transmitted as a continuously varying electrical signal across a pair of wires (a local loop or a subscriber line loop) between the handset of the subscriber (the phone user) and

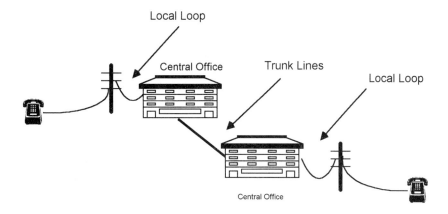

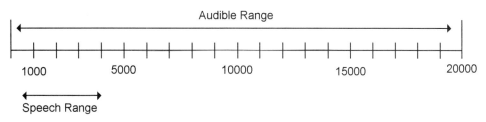

Figure 11-3. Analog Voice Network

an exchange, central office, or end office (See Figure 11-3). Before digital dialing, end offices had names like Prospect or Elgin, so telephone numbers began with an abbreviation of the end office name, for example, PR6-6178 or EL3-1978.

The electrical signal varied in frequency from about 600 to about 3400 Hertz (cycles per second). Although humans can hear frequencies from about 20 Hertz (Hz) to 20,000 Hz, most speech energy is concentrated in the range of 600-3400 Hz.

Trunk lines connected end offices. As the number of end offices grew, it quickly became necessary to organize the phone system into a hierarchy. Too many trunk lines would have been required to interconnect all of the end offices in even a metropolitan area, let alone the whole country (about 20,000 end offices are in use today). By connecting each end office to a toll center, and connecting toll centers together with trunk lines, far fewer trunks were required, yet any subscriber could reach any other subscriber in the area. Over time, a four-layer hierarchy was developed to interconnect all end offices in the U.S.

Long distance trunk lines were expensive to build, yet even the simple copper wires used for the first long distance lines were capable of carrying an electrical signal with a much wider bandwidth than was required for voice transmission. A technique called Frequency Division

Multiplexing (FDM) was developed that allowed a trunk line with only two pairs of wires to carry many voice conversations (many voice channels) simultaneously. Because of limitations in the technology first developed for FDM, a voice channel was standardized at 4000 Hz of bandwidth. The additional 600 Hz between 3400 Hz and 4000 Hz was necessary to separate the channels on a multiplexed line, so that one channel would not interfere with the next.

Telecommunications Components

This section reviews common telecommunications components found in virtually all networks in use today. The intent of this section is to familiarize you with the devices and options in networks which use telecommunications facilities.

DTE/Channel Service Unit (CSU) Interface

DTEs (Data Terminal Equipment) are typically provided with one or more RS-232 "ports" so that they can connect to modems (their manufacturers cannot assume that digital facilities are always available). It makes sense that the same ports would be used to connect the DTE to digital facilities, rather than equipping each DTE with a separate connector of some sort. It follows that something must lie between the DTE and the digital facility, because the digital facility does not use the RS-232 protocol.

The local loop for digital service always terminates at a Channel Service Unit (CSU) in the subscribers building (See Figure 11-4). The CSU is the device that actually generates the transmission signals on the local loop. The DTE connects to the CSU in one of three ways:

1. Through a Data Service Unit (DSU). When digital services were first made available to subscribers, the telephone companies would never allow subscriber equipment to attach

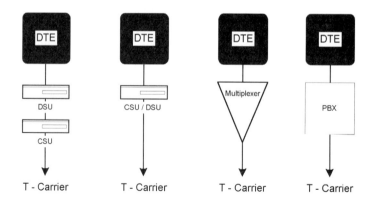

Figure 11-4. DTEs, CSUs, and DSUs

directly to a local loop, so the CSU was provided by the telephone company. The DSU was a separate device provided by the subscriber. Today, CSUs and DSUs are combined in a single device (a DSU/CSU) that is typically owned by the subscriber.

2. Through a multiplexer. We will discuss these devices later in this section.

3. Through a channel bank that is part of a PABX. The CSU will often be built into the PABX.

Multiplexers

A multiplexer is a piece of computer/telephony equipment which allows multiple signals to travel over the same physical media. There are different types of multiplexers such as Time Division Multiplexers and Frequency Division Multiplexers. Figure 11-5 shows three analog voice signals converted by CODECs which are multiplexed onto a serial digital bit stream.

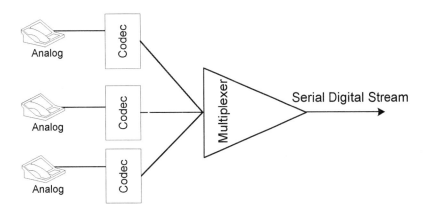

Figure 11-5. Multiplexers

TDM stands for Time Division Multiplexing. This type of multiplexing combines many digital bit streams with relatively low bit frequency into a single bit stream with a relatively high bit frequency. It is, in essence, a way for many slow communications channels to "time share" a very fast channel. The advantage of course, is that the cost per bit transmitted on a single fast channel is lower than on slower channels (See Figure 11-6).

TDM is accomplished by simply interleaving data from several bit streams. This can be done on a bit basis or on a byte basis (this is called bit interleaving and byte interleaving). During time interval 1, 8 bits from source channel 1 are transmitted. During successive intervals, bytes from successive source channels are transmitted on the output channel. A complete set of values from each input channel is called a frame.

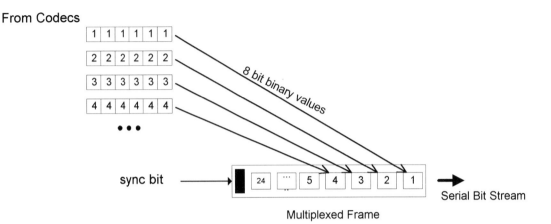

Figure 11-6. Time Division Multiplexers

Modems

The term modem is a contraction of "modulator-demodulator" (modems are called data sets by AT&T). Modems are used in pairs, one at each end of the telephone line (See Figure 11-7). Each modem attaches to the computer or terminal via an RS-232 cable. To transmit a message, the modem accepts binary data from the computer or terminal on the RS-232 interface described below. It modulates the telephone line by generating a signal with an audible frequency. The receiving modem demodulates the signal, generating binary data that are transmitted to the terminal or computer. The signal falls within the range of 300-3400 Hz, so it can be transmitted across the telephone network as if it were a voice conversation.

The simplest protocol represents zeros and ones by switching the audible tone off and on, but that technique can transmit data at a maximum of only 1200 bits per second (bps). More complex protocols can increase the effective data rate by a factor of more than twenty.

The modems at either end of a connection must, of course, use the same modulation/demodulation protocol. A variety of standards have been published by ITU-T and ISO, so that modems from different manufacturers can be used together.

RS-232

RS-232 defines the interface between the computer and the modem. We use the term RS-232 to refer generically to a family of standards developed by the Electronics Industries Association (EIA), ITU-T, and ISO to provide for computer interfaces to the outside world, especially to the

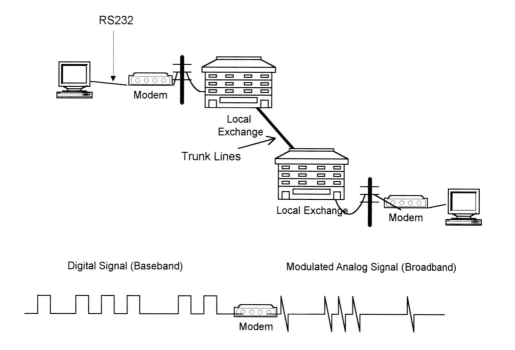

RS232

Modem

Local
Exchange

Trunk Lines

Local Exchange

Modem

Digital Signal (Baseband)

Modulated Analog Signal (Broadband)

Modem

Figure 11-7. RS-232 and Modems

telephone network. Today, the official name for the base standard is EIA-232, but RS-232 is commonly used. These standards define a serial interface (the computer transmits a bit at a time) for data transfer.

Modem Protocols

The DCEs at either end of the communication channel must, share the same protocol. A variety of protocols are defined and in some cases standardized for communication over analog links (between modems).

Any protocol for communication between modems defines:

- The electrical characteristics of the link that determine the maximum bit rate at which the modems can exchange data

- How the modems equalize, that is, how they establish a common voltage reference level

- Whether they operate in asynchronous (start-stop) mode, synchronous mode, or either of the two

- Whether they operate in full duplex or half-duplex mode

Electrical Characteristics. Examples of these characteristics are signal amplitudes, carrier frequencies, and the method used to encode data bits on the channel. It is beyond the scope of this book to explain these in detail. We will simply say that these methods involve modulating one or more carrier frequencies and varying the frequency, amplitude, or phase of the modulating signal.

Equalization. Equalization is the process of establishing a common reference voltage between the modems. It compensates for differences in signal attenuation at varying frequencies. It is required for accurate transmission of data at higher rates.

Asynchronous/Synchronous. Asynchronous operation means simply that bits are not transmitted on any strict timetable (See Figure 11-8). The start of each character is indicated by transmitting a start bit. After the final bit of the character is transmitted, a stop bit is sent, indicating the end of the character. The modems must stay in synchronization only for the length of time that it takes to transmit the 8 bits. If their clocks are slightly out of synch, data transfer will still be successful.

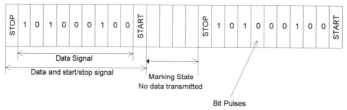

A) Asynchronous transmission of 8 bit data example. The length of each bit pulse is dependent on the transmission speed. If you are transmitting at 300 bits/second, each pulse would be 3.3 microseconds long 1/300).

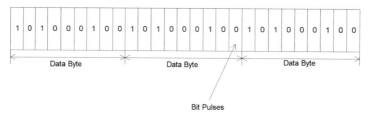

B) Synchronous transmission is typically used for a block of characters, such as frames or packets. Both the sending and receiving devices are operated simultaneously, and they are resynchronized for each block of data.

Figure 11-8. Asynchronous and Synchronous Transmission

Synchronous operation is, of course, the opposite. The modems must first closely synchronize their internal timing circuits, usually by transmitting a burst of bits of a feed length when the connection is established. To transmit data, the sending modem puts a one or a zero on the line every so often. The receiving modem samples the line on the same timetable and transmits the condition of the line (one or zero) to the DTE. They must stay in synchronization in order to communicate.

Full-Duplex/Half-Duplex. A leased line typically has four wires, although it is possible to lease a two-wire line. Half-duplex means that one wire-pair is used to transmit in one direction and the other in another, but transmission does not take place in both directions at the same time. Half duplex is faster than using a single wire-pair for both directions because it is not necessary for the modems to wait for the line to "turn around" each time the direction of data transmission reverses.

Full-duplex means that data are transmitted in both directions at the same time, using one wire-pair for each direction. In principle, full duplex allows data to flow in both directions simultaneously. As a practical matter, the primary advantage to full duplex transmission is that while messages flow in one direction, "acknowledgments" of previously transmitted messages can flow back to the sender in the opposite direction.

It is also possible to implement full-duplex operation on a single pair of wires by using different carrier frequencies in either direction, in a manner similar to FDM.

Private Branch Exchanges

When a large company or other organization had many telephones installed, it became desirable for a variety of reasons to locate a branch of an exchange at the site. These private branch exchanges (PBXs) were leased from the telephone company in the days before deregulation, and were extensions of the same technology used in end offices. With a PBX, callers from outside could dial a single number to reach the organization, and their call would be switched by an operator to the appropriate individual. A PBX can switch internal calls without going through the central office. When a phone call is placed which is destined for another facility, a PBX will route the call to the local exchange (See Figure 11-9).

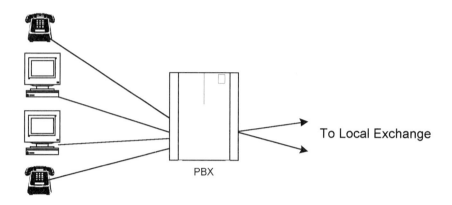

Figure 11-9. Private Branch Exchange

WAN Link Options

Not only are components such as bridges, routers, CSU/DSUs, and modems needed to create MANs and WANs, but links must be purchased to connect the remote sites. (See Figure 11-10).

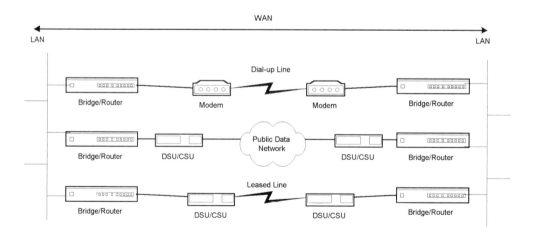

Figure 11-10. Typical WAN Link

There are two different ways to interconnect LANs. The most basic is via a point-to-point network and the second is through a packet-switched network such as defined by the X.25 standard. Point-to-point links establish a physical connection between end points. These links come in a wide variety of data rates as well as costs. These are summarized in Figure 11-11.

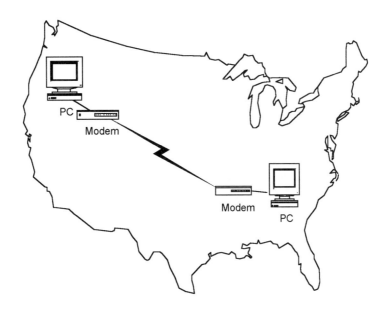

Service	Link Speed	Equipment Needs	Equipment Costs	Monthly Link Costs (Distance Dependent)
Switched Analog	300 bps - 28.8 bps	Modems	$75 - $750	$50 - $750
Leased Analog	300 bps - 28.8 bps	Modems	$75 - $750	$50 - $750
DDS	2.4 Kbps - 56 Kbps	DSU and CSU	$75 - $750	$50 - $500
Fractional T1	64Kbps - 1.544 Mbps	DSU and CSU	$500 - $7500	$150 - $2,500
T1	1.544 Mbps	Multiplexer	$500 - $7500	$300 - $2,500
T3	44.736 Mbps	Multiplexer	$2000 - $20,000	$2,500 - $10,000

Figure 11-11. Point-to-Point Alternatives

Switched services fall into two broad categories, circuit-switched networks and packet-switched networks. Switched services are used when you require more flexibility in connecting WAN nodes. Dialup or dedicated connections are still necessary to reach the switched network, but once the switched network is accessed, you have any-to-any connectivity. Figure 11-12 provides a summary of some of the switched services offered in networks today.

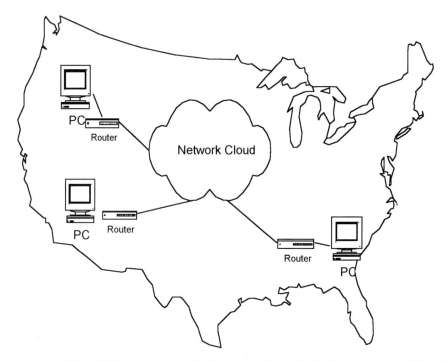

Service	Link Speed	Equpment Needs	Equipment Costs	Link Costs
Switched 56	56 Kbps	CSU/DSU	$100 - $750	$75 - $500
X.25	300 bps - 2Mbps	Interface Card	$100 - $750	$50 - $500
Frame Relay	56Kbps - 1.544 Mbps	Interface Card	$200 - $750	$300 - $2500
ATM	100 - 2.48Gbps	Interface Card	$500 - $1000 per card	$2500 - $10,000

Figure 11-12. Switched Link Alternative

POINT-TO-POINT ALTERNATIVES

Dialup Connections

A dialup line is a circuit that exists between two nodes that uses the switched telephone network to communicate. Dialup lines provide the following capabilities:

*   300—28.8 Kbps Transfer Rates
*   Any-to-any connectivity (one at a time)
*   Requires compatible modems at each end
*   Inexpensive
*   Call initiation must take place before communication occurs

Leased Lines

The telephone system was designed for voice communications (See Figure 11-13). Some errors can be introduced into a voice signal without causing problems for the people at either end of the conversation. For example, a shift in the phase of the signal (essentially a delay in the transmission of the signal) will have little or no effect on voice quality. But such errors can cause problems for data transmission.

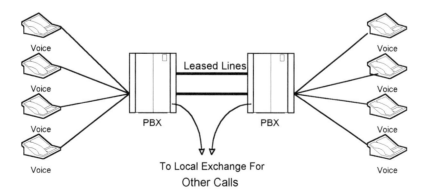

Figure 11-13. Leased Lines

When the telephone network was an analog network, electro-mechanical switches could inject a certain amount of noise into circuits. Other instruments in the transmission path, such as multiplexers, could corrupt the signal further. At lower data rates, say 300 or 1200 bps, modems can correctly transmit data with few, if any, errors even when the line is noisy, but as data rates increase, noise causes more problems. As a practical matter, in the analog network, data could not be transmitted reliably over long distances on switched lines at rates over 4800 bps. Therefore, it became common practice for the telephone companies to "lease" lines to companies for continuous, unswitched use. A company with two computer sites could, for example, lease a line or a set of lines to interconnect the sites.

Leased lines could be "conditioned" by the telephone company to make them usable at higher rates of transmission, up to 19.2 bps. That basic limit was imposed by the previously mentioned 4000 Hz bandwidth of voice channels, although recent advances in error correction and data compression techniques have made effective rates significantly beyond that possible. A leased line would usually be composed, not of an actual dedicated set of copper wires, but of a dedicated FDM channel between area offices of the phone company. To obtain effective data transfer rates greater than 19.2 bps, it was then necessary for the subscriber to lease multiple lines and transmit data in parallel.

Leased lines have an additional characteristic that makes them desirable for certain uses: the length of the circuit is known (the telephone company is usually able to provide this information). The length does not usually vary from day to day. Fixed length is important to certain applications that have timing considerations related to actual circuit length, for example, channel extension. (Backup circuits can be leased for an additional charge.)

Subrate Facilities

Any facility that operates at a data rate less than DS0 (64Kbps) is called a subrate facility. AT&T's Dataphone Digital Service (DDS) and British Telcom's KiloStream are examples of subrate facilities. The first digital telecommunications facilities made available to subscribers were based on DS0 in North America and equivalent services in other parts of the world, and were therefore subrate facilities. Other subrate facilities are very similar.

Dataphone Digital Service (DDS)

DDS was first made available in 1974. It provides the subscriber with access to a digital network spanning the U.S. The subscriber uses all or some portion of a DS0 channel. The following DDS services are offered:

2400 bps leased	19.2 Kbps leased
4800 bps leased	56.0 Kbps leased
9600 bps leased	56.0 Kbps switched

The subscriber must connect to the DDS facility through a DSU/CSU. The local loop operates at the bit rate of the service the subscriber has selected; but from the end office onward, the phone company multiplexes it with other DDS lines and voice channels on T-carriers. This includes multiplexing multiple DDS lines onto a single 64 Kbps DS0 channel.

DDS II was introduced in 1988. It operates at the same speeds and in the same manner as DDS but also provides a diagnostic channel for each primary subrate channel. The subscriber can, with newer DSU/CSU equipment, take advantage of the diagnostic channel for nondisruptive testing and network management purposes.

DDS is not available in every end office. There are about 100 cities in the U.S. that offer the service. Subscribers located outside the DDS servicing areas can access DDS in the nearest DDS city across an analog extension, that is, a pair of modems and a special device for synchronizing the modems with the DDS network.

Multiple subrate facilities can be multiplexed into a single DS0 by the subscriber. This is called subrate multiplexing.

Digital Telephony—T1, Fractional T1, and T3

With the exception of the analog voice signal originating at the microphone of the telephone handset and the audio tones that must be reproduced at the speaker in the handset at the other end, a telephone network is binary in nature. Switches are essentially binary, rotary phones generating a series of pulses to represent numbers. Control signals such as the dial tone, busy signal, and ring are binary. Because of the binary nature of telephony, conversion of the analog network to digital was a natural and logical course. In fact, the conversion was essentially complete in the U.S. for a network then serving 180 million phones less than twenty years after the basic technology became available.

When solid state electronics became available in the late 1950s, voice digitization became feasible. We will explain how analog voice signals are digitized in the next section. The advantages of voice digitization are with the signal having only two possible values (0 and 1).

- It is less susceptible to interference and is easier to tell noise from signal.

- It can be reproduced exactly when it passes through transmission equipment.

- It is easier to mix the voice signal with other binary information, such as signals between switches.

Digital signals also made possible a better method of multiplexing, called time division multiplexing. In 1962, the Bell system installed the first "T-carrier" system for multiplexing digitized voice signals. As mentioned above, frequency division multiplexers (FDM) had previously been developed to multiplex analog voice signals. The T-carrier family of systems, which now includes T1, T1C, T1D, T2, T3, T4 (and their European counterparts, E1, E2, etc.), replaced the FDM systems, providing far better transmission quality (See Figure 11-14).

Note, however, that T1 and its successors were designed to multiplex voice communications. Therefore, T1 was designed such that each channel carries a digitized representation of an analog signal that has a bandwidth of 4000 Hz, as we explained earlier. It turns out that 64 Kbps is required to digitize a 4000 Hz voice signal (we explain the digitization process in a later section of this chapter). Current digitization technology has reduced that requirement to 32 Kbps or less, but a T-carrier channel is still 64 Kbps.

T-Carrier rates are shown in the T-Carrier Line diagram. Fractional T1 (FT1) is a service offered by the phone company which provides users of telecommunications services optional data rates from 64Kbps to 1.544Mbps. It is called Fractional T1 because the user can specify the desired rate which is a fraction of the normal T1 rate (1.544Mbps). It is a low cost alternative to purchasing a full T1 and using only a portion of the bandwidth.

	Standard	Line Type	# Voice Ckt	Bit Rate
	DSO	n/a	1	64Kbps
	DS1	T1	24	1.544
North America	DS1C	T1C/D	48	3.152
	DS2	T2	96	6.312
	DS3	T3	672	44.736
	DS4	T4	4032	274.176

	E1	M1	30	2.048
	E2	M2	120	8.448
Europe	E3	M3	480	34.368
	E4	M4	1920	139.264
	E5	M5	7680	565.148

	1	F-1	24	1.544
	2	F-6M	96	6.312
	3	F-32M	480	34.064
Japan	4	F-100M	1440	97.728
	5	F-400M	5760	397.20
	6	F-4.6G	23040	1588.80

Figure 11-14. T-Carrier Rates

T1 circuits are dedicated services connecting networks or LANs over extended distances. Figure 11-15 shows a typical configuration.

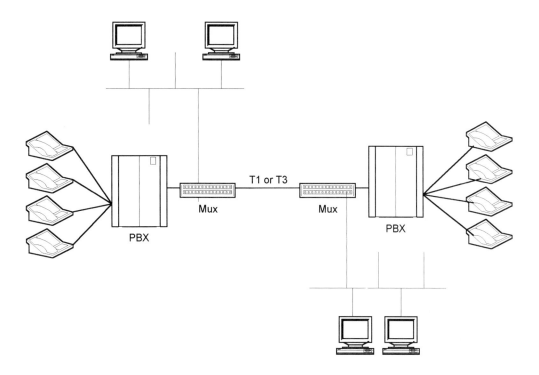

Figure 11-15. Sample T1 Configuration

Switched-56

Switched-56 (SW56) is another service provided by the telephone company (See Figure 11-16). It is digital and requires a CSU/DSU combination to attach a router to the phone line. The interface between the router and the CSU/DSU is typically a V.35 serial cable. Once configured, the switched-56 lines operate as any other dial-out lines by dialing a distance switch-56 station. Switched-56 as often used as a backup line for higher speed connections such as T1.

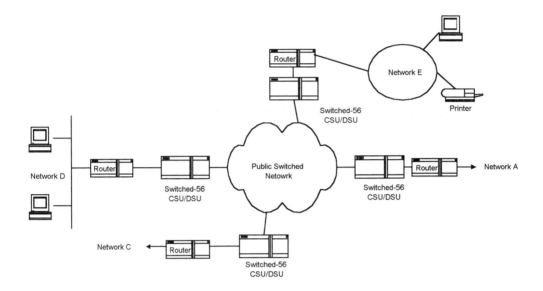

Figure 11-16. Switched-56 Configuration

Synchronous Optical Network (SONET)

SONET stands for Synchronous Optical NETwork. Fiber optics has, of course, been used for some time in the public long distance networks. You may recall Sprint's TV ads where the sound of a pin dropping could be heard over a long distance line. But the links in that first generation of fiber optics were entirely proprietary in nature—architectures, equipment, protocols, formats for multiplex frames, etc. SONET standardizes optical transmission. This has obvious advantages to the telephone companies, making it possible for them to select equipment from multiple vendors and to interface with other phone companies "in the glass," that is, without converting back to copper. SONET will allow synchronous signals as low as DS0 to be switched without being demultiplexed.

The SONET standard defines a signal hierarchy similar to that which you saw for the T-Carriers, but extending to much higher bandwidths (See Figure 11-17). The basic building block is the STS-1 51.84 Mbps signal, chosen to accommodate a DS3 signal. The hierarchy is defined up to STS-48, that is, 48 STS-1 channels for a total of 2,488.32 Mbps-capable of carrying 32,256 voice circuits. The STS designation refers to the interface for electrical signals. The optical signal standards are correspondingly designated OC-1 (Optical Carrier-1), OC-2, etc.

STS and OC	Line Rate (Mbps)	Number DS1s	Number DS3s
1	51.84	28	1
3	155.52	84	3
9	466.56	252	9
12	622.08	336	12
18	933.12	504	18
24	1244.16	572	24
36	1866.24	1008	36
48	2488.32	1344	48

Figure 11-17. SONET Bandwidth

SONET has important advantages for those who use the switched networks for communications:

1. The SONET standards provide a low-level platform upon which standards such as the others described in this section can be based.

2. SONET makes it possible for subscribers to purchase equipment that interfaces "in the glass" with the switched public networks. SONET interfaces are available to SMDS and ISDN for example.

3. Even the earlier offerings, OC-1 to OC-3, make new applications combining data, voice, and video images both technically and economically feasible.

4. The data interface to a SONET network, when SONET links are extended into a subscriber's premises, are through LAP-D at Layer 2. In other words, it looks just like a copper ISDN connection to the subscriber's network.

The SONET standard includes extensive network operation and management facilities. A significant portion of the SONET bandwidth is allocated for out-of-band control signaling for this purpose. This management system has its own OSI-compliant communications architecture. Ultimately, subscribers will be able to interface directly to this capability, running the necessary "stack" on their own computers.

Figure 11-18 illustrates how information is multiplexed onto SONET networks. "Low Speed" signals such as DS1s and DS3 are multiplexed into a single STS-1 bit stream. Three STS-1 bit streams are multiplexed again to form an STS-3 bit stream. In this example, no other multiplexing is performed and the STS-3 signal is converted into its optical equivalent, an OC-3 signal.

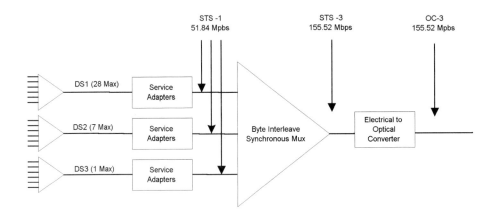

Figure 11-18. SONET Multiplexing

SWITCHING ALTERNATIVES

In this section, you will learn about older switching standards as well as several important emerging standards. In the next few years, these emerging standards will make much greater bandwidth available from the public switched networks for data communications.

The emerging standards we cover in this section are evolving because of these fundamental changes in the telecommunications environment:

1. The greatly improved reliability of the switched public networks as a result of the conversion to digital technology and improved transmission technologies such as optical fiber

2. Today's networks are inherently more reliable than previous analog/copper systems

3. The development of standardized technology that provides much greater bandwidth for transmission over long distances at reasonable cost, namely, fiber optics

4. The emergence of the ability of switched public networks to switch packets of data, rather than just switch circuits, as is the case with the current T-Carrier technology

At the same time, data communications needs are changing. In the recent past, networks were mainframe-oriented, and data communications traffic consisted of a fairly steady stream of relatively small messages. Traffic from dispersed networks was concentrated at the mainframes and was confined to the corporate network because users had little need or ability to communicate with other networks. Building wide area networks linking the mainframes with T1 links was economically justified, because the links were highly utilized most of the time.

Backbones for today's high speed LANs, however, pose different problems. On the one hand, high bandwidth is required because the LANs themselves have very high bandwidth. On the other hand, LAN traffic tends to be very sporadic, with brief spurts of very high activity between LAN nodes being the norm. It is often simply not cost effective to lease T1 links to build a WAN backbone, and LAN users often want to communicate with users on another network outside the enterprise, for example, on the Internet.

In summary, today's users require access to high bandwidth on a dynamic, on-demand basis, rather than on a static, ongoing basis. This implies that the service required is switched. The on-demand packet-switching service provided by the current X.25 networks is simply too slow to serve these needs.

Layer	B-ISDN	Frame Relay	SMDS		MAN
LLC	X.25	I.122	SNI	SIP-3	802.2
MAC	LAP-D			SIP-2 (DQDB)	802.6/ DQDB
Cell	ATM			SIP-1 (ATM)	ATM
Physical	SONET	T1/T3 SONET	T1/T3/SONET		T1/T3/ SONET

Figure 11-19. Telecommunications Technologies

The result of these factors was the meteoric rise in the 1990s of a variety of capabilities involving several new technologies and standards that we summarize in Figure 11-19 . It is organized according to OSI Layers 1 and 2. We have divided the Data Link Layer, OSI Layer 2, into two parts, as we did in The LAN Architectures section: Logical Link Control (LLC) and Media Access Control (MAC). We have again divided Layer 1 into two parts, as we did earlier in this chapter, but this time the division is based on the protocol for the framing of data into cells and the protocol for the physical media.

In the remainder of this section, we will discuss facilities which provide fast, on-demand switched communications now, and will continue to provide over the next several years.

Broadband ISDN (B-ISDN) uses ATM cell relay technology, which is explained in more detail below.

Frame relay, now available to many subscribers, provides switched service based on the switching of frames rather than cells. I.122 is an adaptation of the ITU-T ISDN standards for this purpose. Originally slotted well below B-ISDN performance, frame relay will be further developed by its providers to increase its performance so that it will overlap with B-ISDN at the low end (the T3/E3 range).

Switched Multi-Megabit Data Service (SMDS) is actually a service specification rather than a communications protocol. Its service is defined by the Subscriber Network Interface (SNI) which consists of the SMDS Interface Protocols 1, 2, and 3 (SIP-1, -2, and -3). Like a public X.25 network, SMDS is said "to have no distance," meaning that the subscriber's Network Layer does not see any of the internal nodes of the SMDS network. It uses both cell relay and DQDB.

In addition, we have included IEEE Metropolitan Area Network architecture (MAN), because it shares a great deal of terminology with SMDS.

IEEE has recently defined an extension to the 802 of standards by adding 802.6 at the MAC layer. IEEE 802.6 defines the Distributed Queue Dual Bus (DQDB) architecture. It is similar to FDDI but differs in some important ways.

All of these facilities potentially will use the Synchronous Optical Network (SONET) technology at Layer 1, or alternatively, T1 or T3.

These facilities are differentiated by the speed at which they operate, both currently and planned, whether they provide switched service, the type of data service they offer (connectionless or both connection-oriented and connectionless), and whether or not they provide isochronous service. Isochronous (from the Greek word "one time") applies to connection-oriented services and means that a portion of the channel bandwidth is dedicated to the connection and that synchronous multiplexing is provided. In other words, the channel can be used for voice, video, and other services that require that packets not be buffered or otherwise delayed.

X.25 Overview

Telecommunications in general refers to OSI Layer 1 facilities. Recall that Layer 1 services can be characterized as:

- Bit-stream rather than frame or packet oriented
- Point-to-Point
- Possibly unreliable, because only limited error checking can be performed on a stream of bits

In many cases, the subscriber's applications will indeed interface with the telecommunications facilities at OSI Layer 1. For example, SNA communications across the facilities we have described might use SDLC at Layer 2 and RS-232 at Layer 1.

However, you will recall that another option is to interface to a public packet switching network (or even a private network created by the subscriber that presents the same interface). Refer to Figure 11-20. An X.25 network, whether public or private, is typically built largely upon the facilities of the public telephone networks, in other words, leased lines. In the diagram, connection to the X.25 network is shown through routers which attach remote LANs through a public X.25 facility. An example frame is shown in the format required to access the X.25 network.

In the United States, X.25 is offered by most carriers and VARs. These include AT&T, US Sprint, CompuServe, Ameritech, Pacific Bell, and others.

Several concepts are important for understanding X.25 protocols and networks. These include:

- Physical Circuit: The physical hardware and cabling used to connect devices together

- Virtual Circuit: The complete path between two communicating devices

- Logical Channel: The logical connection between the user node and the network

X.25 has been a long-time standard for packet switching. The X.25 interface lies at OSI Layer 3, rather than at Layer 1. X.25 defines a protocol stack having three layers. We will discuss all three layers.

We cover X.25 at this point in the book for two reasons. First, although the high-level X.25 interface technically falls at Layer 3, X.25 is essentially a telecommunications delivery mechanism. Second, as you will see in the following sections, the public switched networks are gaining the ability to switch packets as well as circuits, and the distinction between telecommunications at Layer 1 and data communications at higher layers is blurring.

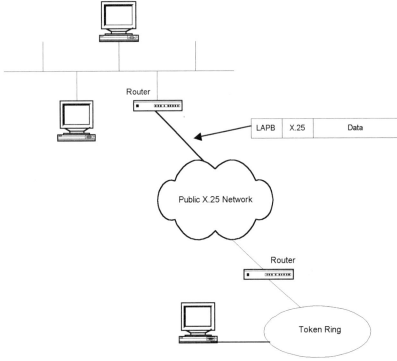

Figure 11-20. X.25 Network

X.25 Services

ITU-T recommendation X.121 defines a system of addressing users that is similar to that used for the voice telephone networks (country code, network code, address). X.25 users anywhere in the world are uniquely addressed, and an X.25 communication can take place between one user and any other X.25 user, as long as the other user is on the same network or a bridge exists between the networks.

X.25 is connection-oriented. Two types of service are offered:

1. Permanent Virtual Circuits. This is the X.25 equivalent of a leased line, statically defined and always available as long as the network is up. Unlike leased lines, however, more than one virtual circuit can share a physical link.

2. Virtual Calls. The network establishes the connection on a virtual circuit, transfers packets until the application is finished, and then releases the connection.

It is important to note that X.25 provides error-free service to the Transport Layer. As you will see later in this chapter, the overhead associated with the necessary error checking is proving to be unacceptable in today's highly reliable digital networks.

X.25 Protocols

Figure 11-21 illustrates the X.25 protocol stack. The X.25 standard predates OSI (the first version was issued in 1976). OSI has adopted X.25 Layer 3 as a connection-oriented network layer protocol.

The X.25 standard does not itself fully define all three layers of the stack, but rather refers to other standards. X.75, for example, is a standard that defines the interface between two distinct X.25 networks and is nearly identical to X.25.

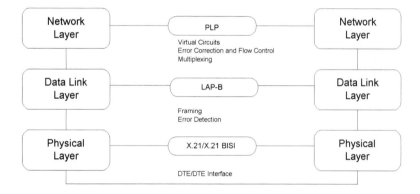

Figure 11-21. X.25 Protocol Layers

X.25 consists of these protocols:

Layer 3-X.25 Packet Layer Protocol (PLP). PLP manages connections between DTEs anywhere in the network, accepting packets from the Transport Layer process and taking responsibility for error-free delivery of the packets to their destination. PLP handles the multiplexing of packets across a link. It establishes virtual circuits and routes packets across those circuits. Many virtual circuits can share a link, so this results in the multiplexing of packets.

Layer 2-Link Access Procedure-Balanced. ITU-T adopted a subset of the ISO HDLC standard. LAP-B is responsible for point-to-point delivery of error-free frames. It is balanced because the LAP-B standard excludes the portions of the HDLC standard that deal with multidrop, "unbalanced" operation.

Layer 1-X.21 and X.21bis. X.21 defines a DTE/DCE interface along the lines of the RS-232 (V.24) standard we discussed earlier, except that it was designed for interfacing to a digital network (such as ISDN). Because digital networks are not generally available, X.21bis, which is essentially RS-232, was defined as an interim standard.

Packet Assemblers/Disassemblers (PADs)

For an application to transmit data across an X.25 network, the application node must have an X.21 or X.21bis interface and must execute processes that provide the LAP-B and X.25 PLP services to the Transport Layer. This precluded network nodes that did not have these components from using the public networks. ITU-T developed a set of standards, informally called the Interactive Terminal Interface (ITI) standards, meant to provide access for terminals and DTEs that cannot execute the layers of X.25. The standards are X.3, X.28, and X.29.

The ITI standards collectively define a "black box", called a packet assembler/disassembler, or PAD. A PAD "assembles" a stream of bytes originating from an asynchronous DTE (for example, from a personal computer) into X.25 packets and transmits them on the X.25 network. Of course, it performs the reverse operations for data sent back to the DTE (See Figure 11-22). To the DTE, the PAD looks like a modem. This means no special software or hardware must be added to the DTE beyond that needed for ordinary asynchronous communications. It is also possible to attach the DTE to the PAD with a point-to-point link using modems.

A PAD can attach several DTEs, performing a concentrator function by placing data from more than one DTE into a packet when possible.

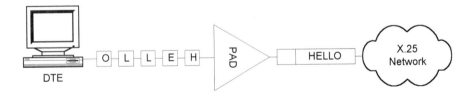

Figure 11-22. Packet Assembler/Disassembler

Overhead and Performance

Since X.25 Packet Layer Protocol (PLP) is responsible for error-free delivery of packets, each X.25 node in a virtual circuit must generate an acknowledgment for each packet received. Meanwhile, the Transport Layer process in the receiving node must also generate an acknowledgment for each packet received (These acknowledgment packets look like data packets to X.25). This means that for every real data packet that traverses the network, three additional packets must also make the trip. The result is that effective throughput is far lower than the rated capacity of the physical links composing the network.

Today's X.25 networks are being replaced by networks employing emerging standards such as Frame Relay.

Frame Relay and Cell Relay: Fast Packet

These standards are called fast packet simply because they are much faster than X.25 or other packet switching methods. This is because they operate at the lowest two layers in the communications protocol and unlike X.25, they can transfer an individual packet at the full bandwidth of today's T-carriers and tomorrow's optical fiber carriers.

Figure 11-23 helps explain the difference between packet switching, frame switching, frame relay, and cell relay. In our earlier discussion of X.25, we showed the division between the enterprise network and the public packet switching network lying between OSI Layers 3 and 4, and we noted that the packet switching networks were "overlaid" on the public telephone networks. Fast packet is moving this boundary down, in the case of cell relay, all the way to the Physical Layer.

When switching is moved below Layer 3, hop-by-hop error recovery and flow control are typically eliminated. The elimination of the Layer 3 acknowledgment packets can potentially double the throughput of a packet switching network, while simplifying and streamlining the network overall.

Both frame relay and cell relay multiplex data from many users over a link, dynamically allocating bandwidth on demand (explained later). This accommodates the "burstiness" of data communications traffic. The relatively small number of virtual circuits that are active at any given moment occupy the full bandwidth of the channel, so very high transfer rates are achieved under conditions of normal load.

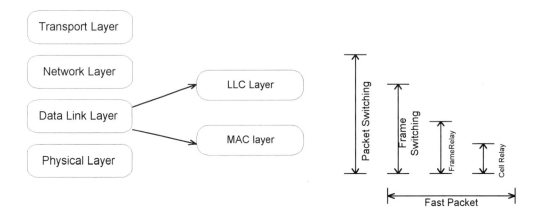

Figure 11-23. Switching and the OSI Model

Integrated Services Digital Network

Having discussed the important telecommunications facilities of yesterday and today, we turn, for the remainder of this book, to the future. We will discuss ISDN first, because, although it will take many more years for its full implementation and more years after that for it to realize its full potential, it is here today.

As the analog transmission and switching components were rendered obsolete by superior digital ones, a new set of protocols was needed to allow their full potential to be realized. ISDN provides a framework for the development of these components and protocols. But ISDN's own protocols are a more recent stage in the evolution of the digital network than an impetus for its continuance.

You have already seen how the telephone companies have largely completed the conversion of their voice networks from analog to digital. You have also seen how they have made the resulting integrated digital network (IDN) directly accessible for data communications by first offering DDS and later, T-Carrier Service in North America and equivalent services elsewhere. You have also seen the limitations of the services offered. ISDN represents a logical migration of the voice oriented IDN toward a network that serves multiple purposes—voice, data, video, facsimile, and all other forms of electronic communication, regardless of the source.

ISDN can be characterized in two ways:

- As a bundle of services offered for the transmission of voice, data, and other forms of communication via the switched telephone networks of the world

- As a set of protocols that defines a standard interface to the network, allowing many vendors to supply both hardware and software that take advantage of the services offered

ISDN Services

You subscribe to ISDN in the same way that you subscribe to voice services. The end office you connect to must, of course, offer ISDN. The process of making ISDN available is a gradual one, so only some end offices currently offer it.

We mentioned earlier that the public switched networks have begun to adopt common channel signaling (CCS) through the SS-7 signaling protocol. ISDN takes advantage of this capability. All signaling for ISDN is out-of-band, that is, CCS.

ISDN provides access to digital channels. Three types of channels are available for subscriber use (See Figure 11-24):

Channel	Rate (Kbps)	Applications
D	16	Control Signaling
B	64	Data Voice Facsimile Slow-scan video
HO H1 H2	384 1536 1920	Backbone networks Full-motion video Multiplexing

Figure 11-24. ISDN Access

D Channels. These channels operate at 16 Kbps and are provided for CCS but can also be used for data. Each D channel is associated with one or more channels of another type. It can be used, for example, to tell the telephone company to which ISDN subscriber the other channels are to be connected. CCS eliminates the problem of distinguishing signals from data. By using a single channel for signaling for several data channels, bandwidth is saved. The D channel can also be used for transmitting certain types of data that require low bit rates.

B Channels. These channels operate at 64 Kbps and are used for data, voice, facsimile, slow-scan video, and so on. Slow scan video refers to video applications which do not require smooth motion of pictures, for example, transmitting the slides of a presentation.

H Channels. These channels operate at 384 (HO), 1536 (HI), or 1920 (H2) Kbps and are used for applications requiring high bandwidth, such as backbone networks and full-motion video. They can also be multiplexed by the subscriber in the same manner as T-carrier channels.

Note that the basic building block for ISDN channels is DS0: B channel = 1 x DS0, HO channel = 6 x DS0, H1= 24 x DS0, and H2 = 30 x DS0. Two basic services are offered from which you may choose, depending on your needs. Of course, you might require more than one kind of service, just as you might require more than one phone line.

Basic Rate Service provides one D channel and two B channels (This service is therefore sometimes referred to as "two B plus D" or "2B+D"). Although it provides 144 Kbps of usable capacity, framing, synchronization, and other overhead bring the total bit rate of basic rate service to 192 Kbps.

Primary Rate Service is structured around the bandwidths of T1 for North America (1.5441544mbps Mbps) and El elsewhere (2.048 Mbps), including Japan. It includes an optional D channel and some number of B and H channels, in combinations that do not exceed the allowable bandwidth when necessary overhead is included.

Having selected the services you require, you can set up connections for the channels in several ways:

- Semi-Permanent. Set up by prior arrangement, this is the ISDN equivalent of a leased line.

- Circuit-Switched. This is similar to using a modem on today's switched public network to establish a connection to another user. An important difference is that the D channel is used to transmit the control information necessary to establish and terminate the call.

- Packet-Switches. This is X.25 packet switching. Software is now available that will let you access ISDN X.25 networks as if they were ordinary X.25 networks.

ISDN also provides a number of services not previously available to data communications users. Many of these are similar to services provided to users of the voice network. For example, ISDN can provide the subscriber number for incoming calls, block incoming calls, transfer a call to another ISDN subscriber, and connect to multiple ISDN subscribers.

ISDN Protocols

ISDN is defined by a set of standards written by ITU-T. The standards are called the I-series of recommendations, first issued in 1984 and updated in every four years. There are seventy-five standards documents in six sections: general ISDN structure, service capabilities, overall network aspects and functions, ISDN user-network interfaces, internetwork interfaces, and maintenance principles. It is clearly far beyond the scope of this book to describe these standards in any detail, but we will describe the key features of the ISDN user-network interface and some of the important communications protocols that have been adopted.

Because ISDN uses common channel signaling (on the D Channel), there are two sets of protocols, one for CCS and one for data (channels B and H).

Physical Attachment to ISDN

A connector similar to that used to connect a telephone handset is defined. The cable connecting the DTE to ISDN has eight wires.

Physical Layer-Frame Formats, I.430, I.43.
These standards describe the protocol used at the Physical Layer (OSI Layer 1) to connect to ISDN. They define the multiplexed frame format for the basic and primary service interfaces respectively.

Data Link Layer-Call Setup, LAPD.
Link Access Procedure-D (for "D channel") defines the protocol used on the D channel to interface with the phone company's SS7 network for setting up calls and other signaling functions.

Data Link Layer-LAPB.
The Link Access Protocol-Balanced defined for X.25 serves as the protocol for the B channel for packet switching in ISDN.

Network Layer-Call Control, I.451.
I.451 defines a high-level protocol to control switching and other signaling to SS-7.

Network Layer-Packet Switching, X.25.
ISDN defines X.25 as the protocol used at the Network Layer for packet switching.

ISDN User Premises Functions.
ISDN defines a set of terms to describe the various pieces of equipment that will be found on the subscriber's premises. By defining this equipment, they further define the interface to ISDN (See Figure 11-25).

1. TE1 (Terminal Equipment type 1). This refers to DTEs that support the I.430 ISDN framing protocol.

2. TE2 (Terminal Equipment type 2). If a DTE is not a TE1, it is a TE2 and requires a terminal adapter to attach to ISDN.

3. TA (Terminal Adapter). TAs are used to attach a TE2 to an NT1 or NT2.

4. NT1 (Network Termination 1). This device terminates one ISDN local loop at the subscriber's premises and attaches TE1s or TE2s (through TAs). It supports multiple channels, multiplexing them onto the ISDN local loop. It operates at OSI Layer 1.

5. NT2 (Network Termination 2). This is an intelligent device operating at the Network Layer (OSI Layer 3). It attaches to the ISDN local loop through an NT1. It can perform switching and multiplexing of TE1s/TE2s.

6. NT12 combines the functions of NT1 and NT2 in a single cabinet.

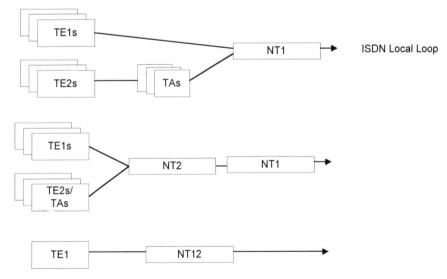

Figure 11-25. ISDN Attachment

Broadband ISDN (B-ISDN) and Asynchronous Transfer Mode (ATM)

Knowing that subscribers can today acquire T3/E3 carriers that provide several times the bandwidth of T1/E1, it is not surprising that T3/E3 bandwidth can be used for ISDN. B-ISDN goes far beyond just providing rates greater than the primary rate.

First, B-ISDN defines a number of additional services, the most interesting of which is video conversational services, including video telephony. Other services are messaging (store and forward) for e-mail and retrieval services for accessing databases.

A second important aspect of B-ISDN is that Asynchronous Transfer Mode (ATM) is specified as the target solution to replace X.25 as the interface between data communication users and ISDN. ATM will provide fast, streamlined packet switching. This implies that B-ISDN will be a packet-based network, or at least present the appearance of such. We will discuss ATM further in the following sections.

Frame Relay

Frame Relay is defined by the ITU-T recommendation I.122 as an adaptation of the ISDN interface for the purpose of providing frame relay services. Like X.25, the standard defines an interface between the enterprise network and the packet-switching network (See Figure 11-26).

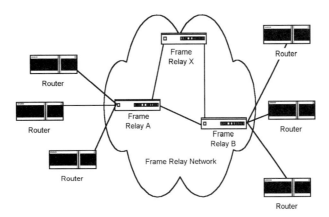

Figure 11-26. Frame Relay Network Example

The term frame is used because frame relay performs asynchronous multiplexing of frame data from multiple virtual circuits. The term relay is used because slots within the frame are not demultiplexed. Several long distance carriers offer frame relay. In the Frame Relay service, LAPD is stripped to its bare essentials and then stripped again. Frame Relay is based on the core aspects of LAPD protocol (ANSI T1.618), an extended subset of LAPD. Call control is typically on the D channel, using a form of DSS1 specified in CCITT Q.933 and ANSI T1.617. Frame Relay is based on the idea that LAPD per se isn not really what data networks want or need for an efficient packet switched service. Frame Relay is even simpler, at least in terms of visible protocol elements (it uses fewer header bits). Some proponents consider it to be close to circuit switching.

A frame relay is essentially an electronic switch. Physically, it is a box which connects to three or more high-speed links and routes data traffic between them. The Frame Relay diagram illustrates its operation. Suppose virtual circuits have been established as shown in Figure 11-27: Virtual circuit 1 from multiplexer A to multiplexer C, virtual circuit 2 from multiplexer A to multiplexer D, and virtual circuit 3 from multiplexer A to multiplexer E. All three circuits flow through Frame Relay B.

Next, suppose data for all three circuits flow into multiplexer A. The MUX places them into the frame, storing an address and length with the data (the diagram is simplified by showing all data the same length). The frame is transmitted from A to B. B must demultiplex it and then create frames for C, D, and E.

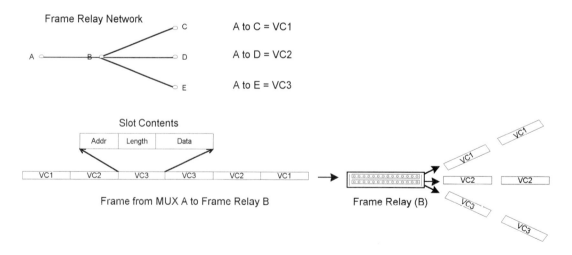

Figure 11-27. Frame Relay

Frame relay has these characteristics:

- It is defined by ITU-T, T1, or E1 bandwidth. It is the ITU-T's intention that Frame Relay fall under cell relay (ATM) with respect to satisfying user bandwidth requirements. However, current Frame Relay service offerers are promising T3 bandwidth, so Frame Relay will overlap with cell relay at its high end.

- Frame Relay is intended only for data communications, not voice or video, for the same reason that Stat MUXes are limited in this way.

- Only connection-oriented service is provided. Unlike other ISDN packet services, the network does not offer a complete connection mode data link service. While Frame Relay is connection oriented, a frame either makes it across the network or it doesn't. (Its address field, and thus the CRC, are local to each interface and are changed by the network).

- Transmission errors are detected but not corrected (the frame is discarded). The sender at the DLC level is not notified that the packet was discarded. Frame Relay facilities are assumed to have low bit error rates.

- Frame Relay is faster than X.25 because only the data link layer is needed and no error correction is performed within the network.

- Frame Relay is currently offered by most of the major carriers such as AT&T, MCI, and Sprint. It is not suited for voice and video because of its variable length frames.

Frame Relay adapts to a wide variety of traffic. Figure 11-28 shows the type of traffic that is intended to be used in a frame relay frame. Note that all of these examples are data and not voice or other time sensitive information. Frame relay is well-suited for data either at the Data Link Layer or the Network Layer. Examples of Data Link Layer protocols supported by Frame Relay are Ethernet and Token Ring. Network Layer protocol examples are IP and X.25. Frame Relay can handle multiple layers of data communications protocols, such as SNA datastream, as well.

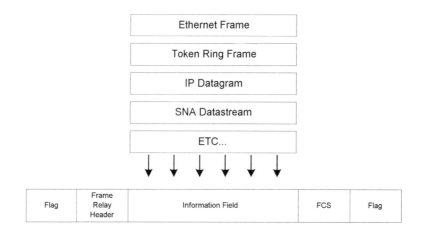

Figure 11-28. Possible Frame Relay Contents

Cell Relay, B-ISDN, and ATM

Cell Relay simplifies packet switching even further. It lies at OSI Layers 1 and 2 and uses a small, fixed-length cell. It is used in the interface to SMDS (discussed below), in B-ISDN. The first ATM switch, which operates at 622 Mbps, was installed in late 1991 and has been implemented in test environments and in a handful of corporations.

Cell Relay achieves the same goal as Frame Relay—asynchronous multiplexing in a fundamentally different way. This is best illustrated graphically (See Figure 11-29).

Making the same assumptions as the previous example (Frame Relay section), the Cell Relay A creates cells from each source. The sources may be data, video, voice, or any other source of digital data. Each 53-byte cell contains 48 bytes of user information and a 5-byte header that simply indicates to which virtual circuit the cell belongs. Obviously, only data from that circuit can go in that cell. This is much simpler than the multiplexing required for Frame Relay.

Cell Relay B relays each cell, a cell at a time, to the channel assigned to the virtual circuit indicated by the header of the cell. Again, this is a very simple operation, no demultiplexing of the input is required.

If you compare the sequence of bits that flows across the communication channel for these examples, you will see that Cell Relay achieves the same multiplexing effect as Frame Relay. However, its operation is much simpler and, therefore, faster and more efficient.

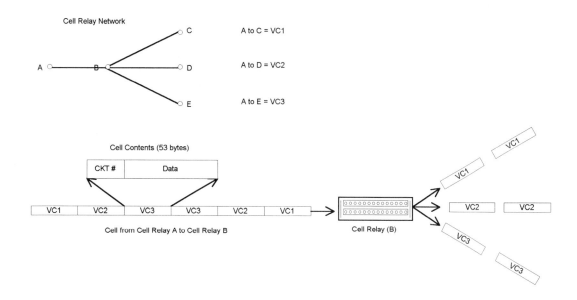

Figure 11-29. Cell Relay

Asynchronous Transfer Mode

Asynchronous Transfer Mode (ATM) is a technology that can transmit voice, video, and data across local area networks (LANs), metropolitan area networks (MANs), and wide area networks (WANs). It is an international standard defined by ANSI and ITU-TSS that implements a high-speed, connection-oriented, cell-switching technology, designed to provide users with scalable bandwidth.

In the mid-1980s, researchers began to investigate the technologies that would serve as the basis for the next generation of high-speed voice, video, and data networks. The researchers took an approach that would take advantage of the anticipated advances in technology. The result of this research was the development of B-ISDN of which ATM was an integral part.

ATM is a connection oriented networking technology that transports data in fixed length cells. ATM has the potential to solve many of today's network problems. It can unite voice, data, images, and video as well as link LAN and WAN networks.

ATM offers a solution to the requirement for high bandwidth and multimedia transmission. It is adaptable to high switching-speeds to support large amounts of bandwidth, and it accomodates traffic of any medium. The term asynchronous in ATM means that cells are transported through a network without requiring the cells to occupy specific time slots in a frame alignment, such as required for T1 frames. Although the individual bits within a cell are synchronized, the timing between cells can vary.

Since the link between end-nodes does not provide a permanent circuit between an information source and its destination (as in circuit switching), it is said to provide a virtual circuit or a circuit that remains for the time it takes the cell to pass through the network. This concept has its roots in X.25 packet-switching. There are two types of virtual circuits: permanent virtual circuits (PVCs) and switched virtual circuits (SVCs).

A PVC behaves like a dedicated line between source and destination end-points: when activated, a PVC will always establish a path between these two end-points.

SVC is analogous to public-switched telephone service: calls can be made dynamically between a source end-point and any destination end-point in the network.

Why the Need for ATM?

The ever-increasing demand for bandwidth has caused the network industry to develop a method that is capable of transferring various types of data at high speeds. The problems facing these users are:

- Support of enterprise-wide work groups on dissimilar computing equipment
- Support of network devices operating on multiple protocols
- Different network topologies, technologies and protocols
- The need to integrate local- and wide-area networks to communicate information over the enterprise

Users are seeking long-term solutions that will not be made obsolete by a better solution within a short time period. They are looking to the telecommunication industry and to private network vendors for direction. Over the past several years, the telecommunication industry has been evaluating transport technologies for proposed broadband services beginning at the DS3 level (45 Mbps) and extending upward. These broadband services would meet three essential criteria:

- Provide expansive, scalable bandwidth capacity
- Support all forms of communication media
- Provide a common transport mechanism for both LANs and WANs

The Benefits of ATM Technology

When applied to LAN internetworking, ATM offers all of the benefits that are ascribed to frame relay—e.g., the ability to statistically multiplex bursty LAN traffic and provide bandwidth on demand. However, there are key differences between the two. Unlike frame relay, ATM can transport and switch both packet-oriented (LAN) and time senstitive (voice, video, and image) traffic. Also, switching and transport speeds associated with ATM are higher than frame relay.

ATM Bandwidth Scalability

ATM offers a scalable approach to the rising demand for more LAN/WAN bandwidth. ATM switch fabrics can be designed for full bandwidth, nonblocking operation. Then, more switching capacity can be achieved by cascading switches to form higher-capacity switch fabrics. Scalability enables users to utilize the amount of bandwidth needed to satisfy their existing needs. As requirements for additional bandwidth emerge, more bandwidth can be supplied. ATM technology is scalable over a spectrum of multimegabit rates, providing virtually unlimited availability of bandwidth.

There is increased emphasis to support very high data rates over twisted pair copper wire, since twisted pair wiring is ubiquitous and is much less expensive than fiber and fiber-optic components. The increasing demand for higher data rates over copper was the impetus for the American National Standards Institute (ANSI) to develop a standard for FDDI for copper, called the Copper Distributed Data Interface (CDDI).

Current WANs can give rates of up to T3, but at great expense. Before the emergence of LANs, requirements for increased bandwidth applied only to WANs. Now LAN users need to be concerned with the bandwidth available to satisfy their LAN applications, and they also need to be concerned about the availability of WAN bandwidth to satisfy their internetworking needs.

The emergence of LAN internetworking introduced new bandwidth limitations. The bandwidth of available WAN transmission facilities did not match LAN bandwidth, thereby introducing greatly reduced performance. WAN digital transmission services have been much improved over the past decade and now offer rates up to T3/E3, but at great expense. However, the growing trend toward rising WAN backbone and internetworking traffic, from multimedia and distributed computing applications throughout an enterprise network, will require carrier services with capacities beyond those of existing services.

ATM Deployment in Wide-Area Networks

The presence of ATM in wide-area networks (WANs) will surface initially in the form of SMDS connectionless public network service. SMDS is both a metropolitan-area and a wide-area service offered by the major common carriers.

ATM networks are characterized by their lack of error protection and flow control, connection-oriented mode operation, reduced header functionality, and small information field. These characteristics reduce transmission delay from one node to another.

Error Protection and Flow Control

ATM does not employ error protection or flow control. Packet loss due to queue overflow is another problem for ATM since there is no flow control provided between nodes. If an error occurs during transmission, no special action is taken to correct the error. Errors are relatively few due to the very high quality of network hardware and the connection-oriented mode of ATM.

Connection-Oriented Mode of ATM

ATM is a connection-oriented network; before any information is transferred, a connection must be established first. During the connection set up phase, the network checks to see if enough resources are available for the connection. If sufficient resources are available, they are reserved and a virtual connection is established, otherwise the connection is refused. When the transfer phase is completed, resources are released. The reservation of resources and the establishment of a virtual connection reduces overflow errors and allows for reduced header functionality.

Reduced Header Functionality

ATM headers have limited functionality allowing for very fast processing and reduced delay. The basic function of ATM headers is for the identification of a virtual connection which is established at call set-up and allows for the proper routing of each packet. The limited functionality of the ATM header allows for simple processing at very high speeds by ATM nodes.

Small Information Field

ATM packets have small information fields. The small size of the ATM information field (48 bytes) allows ATM nodes to have smaller internal buffers, thereby reducing queuing delays.

The broadband network of the future will have to provide for a wide variety of transfer services. These services include low speed data, voice, telefax, hifi sound, and video. The transfer mode of the future must be capable of efficiently providing for all of these services. In order to see how ATM fulfills these needs, an understanding of the basic principles of ATM is needed.

Information Transfer

ATM is based on asynchronous time division multiplexing and the use of fixed length cells. Each cell contains a header and information field. The primary purpose of the header is to identify which virtual pathway or channel a cell belongs to, and to provide the appropriate routing. Cell sequence is preserved per virtual channel. The information field contains the data being transported through the network. No processing is performed on the information field as it travels through the network.

All services (voice, data, video, etc.) can be transported via ATM. To provide for these services, several ATM Adaptation Layer (AAL) processes have been defined. The information field of an ATM cell may contain information used by ATM depending on the AAL process being used.

ATM Routing

ATM is a connection-oriented network. Header information is assigned for each section of a connection for the complete duration of the connection and translated when switched from one section to another. Two types of connections are possible: virtual channel connections (VCC) and virtual path connections (VPC). Virtual path connections are a grouping of VCCs. When switching is performed on cells, it is first performed on the VPC, and then (if necessary) on the VCC. An example of VCC and VPC use might be video conferencing, where video and voice are transported over two VCCs and grouped into one VPC. The VPI and VCI fields (virtual path identifier and virtual channel identifier) are used to identify which VPC and VCC a cell belongs to. An example of VPI and VCI use is illustrated in Figure 11-30.

ATM Resources

ATM is a connection-oriented network where connections are established either semipermanently or for the duration of the call. This includes the allocation of a VCI, possibly VPI, and other resources expressed in terms of throughput (bit rate) and Quality of Service. These resources are negotiated between user and network for switched connections, during the call set-up phase, and possibly during the call.

Proper bandwidth allocation is required for the establishment of VCCs and is important for optimum ATM performance. If too little bandwidth is reserved for a VCI, Quality of Service cannot be guaranteed. Quality of Service (QOS) relates to cell loss and the delay variation of cells belonging to a connection. If too much bandwidth is reserved for a connection, cell loss and delay increase for all connections.

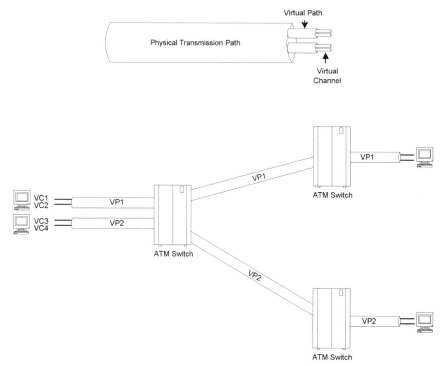

Figure 11-30. ATM Routing

The ATM Cell

Each ATM cell contains a 5-byte header and a 48-byte information field. The ATM header contains six fields (See Figure 11-31). These are described below:

> **GFC** (General Flow Control): 4 bits long and reserved for future flow control use.
> **VPI**: 1 byte long and identifies the virtual pathway a cell belongs to.
> **VCI**: Identifies the virtual channel a cell belongs to. The VCl is a two byte field.
> **PTI** (Payload Type Indicator): 3 bits long and used to indicate the ATM cell type.
> **CLP** (Cell Loss Priority): Indicates the priority for which the cell can be discarded during heavy network congestion.
> **HEC** (Header Error Control): Occupies the last 8 bits of the ATM header. The HEC contains error control information about the header and is used for error detection. If an error is detected, the cell is discarded.

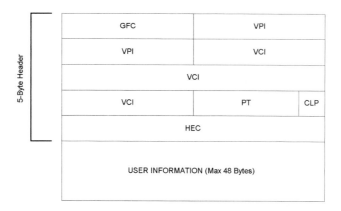

Figure 11-31. The ATM Cell

ATM Connectivity

Within the ATM transport network there are five defined levels of Connectivity. The physical layer contains the lowest two levels: the Transmission Convergence Sublayer, and the Physical Medium Dependent Sublayer. The ATM layer sits on top of the physical layer, and two levels of the ATM Adaptation layer (AAL) rest on the ATM layer. These layers are represented in the BISDN protocol stack pictured in Figure 11-32.

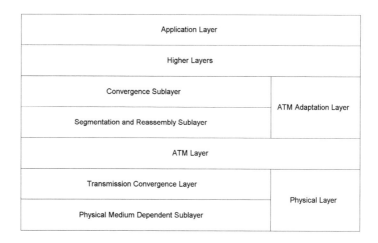

Figure 11-32. BISDN Protocol Stack

ATM Adaptation Layer

The AAL is an end-to-end process used only by the two communicating entities to insert and remove data from the ATM layer. At every point in the network, processing is done only on each ATM cell in isolation. At no point is it necessary to gather more information than is contained in a single cell header to complete the process. The network is not concerned with arrivals of groups of cells, sequencing, or acknowledgment. The AAL enhances the service provided by the ATM layer to a level required by the next highest layer. These needs are met by five different processes:

- AAL 1 is designed for Constant Bit Rate (CBR) data that requires synchronization between the sender and receiver. This service is used by voice, video, and similar traffic. AAL 1 adds 4 bytes of overhead for encoded timing information to address the close coordination needs of CBR traffic.

- AAL 2 addresses the needs of Variable Bit Rate (VBR) services. The additional overhead needs for AAL 2 are not yet defined.

- AAL 3 and 4 are designed for connection and connectionless oriented data services.

- AAL 5, Simple and Efficient Adaptation Layer (SEAL), is used for VBR data transmitted between two users over a preestablished ATM connection. The SEAL AAL assumes that higher layer processes will handle error recovery and is the only AAL process that does not add any overhead to the user information sent to the lower levels. Each AAL process is performed at the Convergence Sublayer (CS).

Convergence Sublayer

The Convergence Sublayer is service dependent. Services provided by the CS include: handling of cell delay variation, source clock recovery, monitoring of lost and misinserted cells and possible corrective action, monitoring of user information for bit errors, and reporting on the status of end-to-end performance. In addition to the services listed above, the CS is also responsible for dividing data from the next higher layer into logical packet data units (CS-PDU) that are usable by the Segmentation and Reassembly Sublayer.

Segmentation and Reassembly Sublayer

The Segmentation and Reassembly Sublayer (SAR) is responsible for the segmentation of higher layer information into 48-byte fields suitable for the payload of ATM cells, and the inverse operation, reassembly of cell payload information for the next higher layer.

ATM Layer

The ATM layer is responsible for data transmission between adjacent nodes and the transportation of data between the AAL and physical layer. During the transportation of data between the AAL and physical layers, the ATM layer adds (or removes) 5 bytes of header information to the SAR-PDU, making a complete ATM cell. In addition to transporting data, the ATM layer is responsible for the multiplexing and demultiplexing of cells into a single cell stream on the physical layer, the translation of the cell identifier when the cell is switched from one physical link to another, and the future implementation of flow control mechanisms supported by the GFC bits.

ATM Physical Layer Interface

The physical layer is made up of two sublayers: the Transmission Convergence and Physical Medium sublayers. The Transmission Convergence sublayer is responsible for converting the ATM cell stream into bits to be transported over the physical medium.

Transmission Convergence Sublayer

The Transmission Convergence (TC) sublayer is responsible for providing the following services: cell rate decoupling, HEC header sequence generation/verification, and cell delineation. To ensure that a useful rate is maintained for the available payload, unassigned cells are either inserted or suppressed. This process is called cell rate decoupling. This sublayer also generates the HEC (Header Error Check) field for each cell at the transmitter and its verification at the receiver. This mechanism also permits the recognition of the cell boundary (cell delineation). The final service provided by the TC is transporting the bit stream to/from the Physical medium.

Physical Medium Dependent Sublayer

This lowest layer on the BISDN protocol stack is the only fully medium-dependent layer. The physical medium is responsible for the correct transmission and reception of bits on the physical medium. This sublayer must guarantee proper bit timing reconstruction at the receiver. Therefore, it is the responsibility of the transmitting peer entity to provide for the insertion of the required bit timing information and line coding.

One of the advantages of ATM is that it will likely not be made obsolete for some time by growing technologies. This is important to many businesses and telecommunications companies who do not wish to keep reinvesting money in a faster network system. With ATM's high transfer rates, hardware and software systems will be working hard just to catch up. ATM starts at DS3 level (45 Mbps) and extends upward from there.

Transmitting real-time voice using packet-oriented technology can be a difficult task because voice cannot tolerate variable delays normally associated with packet traffic. However, ATM supports voice by combining high-speed transmission and very short fixed-length packets with an ATM Adaptation Layer optimized for voice. ATM concurrently supports voice and packet data transmission for use in a multimedia workstation.

Migration from Existing Facilities to ATM

The migration from one technology to another is a major user consideration. This is especially true for the migration from existing LAN technology to ATM. A user must consider changes to the existing infrastructure—such as premise wiring, hardware, and software—and they must be sensitive to the disruption and cost of migration. Users should consider a gradual transition from existing technology to ATM, not a quantum leap which requires significant retrofit. ATM can utilize today's existing cabling infrastructure, running over fiber or copper, at 100 Mbps, or over unshielded-twisted pair (UTP) at 10 or 16 Mbps. It can use existing physical-layer transmission technologies, including those developed for Ethernet, Token Ring, or FDDI/CDDI.

Summary of ATM

ATM is a network protocol capable of providing for current and future telecommunication needs. These needs include high-speed data transfer, high quality videophony, and digital TV. ATM provides for these needs by offering five different processes within the AAL, each offering different services for different needs. In addition, ATM uses small cells, allowing ATM to route packets quickly and with little delay through the network. The layered approach ATM uses allows ATM to meet the probable needs of tomorrow's networks.

Metropolitan Area Networks

IEEE 802.6 and DQDB

As you saw in Chapter 4, FDDI uses token passing to control access to the ring. When a node on an FDDI ring needs to transmit, it must wait until it receives the token, even if the FDDI link between it and the next node in the ring is idle. This amounts to wasted time. An unacceptable amount of time would be wasted if the transmission speed of FDDI were increased beyond its current 100 Mbps specification. DQDB is meant to replace FDDI for MANs with SONET speeds (51.84-2488.32 Mbps). IEEE 802.6 is meant to provide an interface to a DQDB MAN that is consistent with the other 802 protocols (Ethernet, Token Ring, and Token Bus) at the Data Link Layer. 802.6 will allow DQDB bridge nodes to interconnect LANs transparently within an area of at least 50 km (Although the 802.6/SNI for SMDS can potentially become a wide area interconnection).

Dual Buses

You will recall from the LAN Architectures section that FDDI employs dual rings. An FDDI node can connect to either or both rings.

Physically, DQDB looks just like an FDDI ring (ignoring differences in transmission technology and speed), but logically, it is quite different. Even though dual rings are physical, they are logically treated as buses. The term "bus" is used somewhat differently for DQDB than for other technologies, such as Ethernet. Recall that for Ethernet, as for most "bus" architectures, all nodes connected to the "bus" (the Ethernet coax cable) are electrically common. A signal transmitted by one node is immediately received by all other nodes on the bus. This is not the case for the DQDB buses. Each "bus" is composed of several distinct links, like a Token Ring.

A DQDB MAN works like this (See Figure 11-33): On Bus A, messages flow from Node 1, the Head Node, to Node 2. Node 2 relays each message to Node 3, and so on, until finally the messages reach Node n, the tail node. Node n does not simply retransmit; messages "fall off the end," so to speak. Similarly, on Bus B, messages flow from Node n, the head node for that bus to Node n-1 and ultimately to Node 1, the tail, where they likewise fall off the end. Obviously, a node originating a message for a higher numbered node must transmit it on Bus A, and a message for a lower numbered node must be transmitted on Bus B.

Physically, to accommodate the physical ring topology, the head and tail nodes are actually the same node during normal operation, as shown in this figure. Data flow in opposite directions on the buses, so this is called a "counter rotation" scheme.

Self-Healing Operation

A major advantage of counter rotation is recoverability. According to the standard, every node has the ability to take over the function of a head or tail node when a link fails. This allows the MAN to recover from failures in a manner somewhat similar to FDDI. When an FDDI link fails, shunts within the nodes to either side of the failure can "shunt" the dual rings into a single larger ring. With DQDB, if the link between Node 5 and Node 4 fails, Node 4 becomes the head node for Bus A and the tail node for Bus B, while node 3 assumes the opposite roles.

When configured like this, any node can still reach any other node on the MAN, and overall capacity is reduced only by a factor of 1/n (1/5th in this case), because both buses continue to operate at full speed with the remaining four links. With FDDI, by comparison, overall capacity would be reduced by one-half because two rings effectively become one.

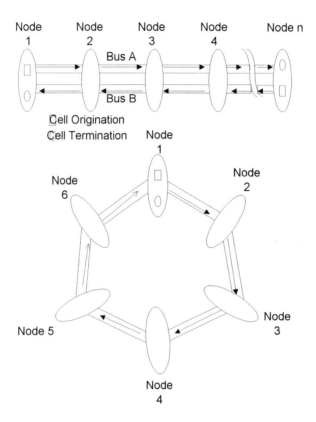

Figure 11-33. DQDB

Distributed Queue

We have explained the "dual bus" portion of the name. It is the "distributed queue" portion of DQDB, however, that allows it to perform at higher rates than FDDI. DQDB uses a scheme called "slotted bus with reservation." The literature refers to DQDB "slots," but we will refer to "cells," since DQDB slots are the same size as ATM cells and will most likely be ATM cells at some future time.

The head node puts cells on the bus, one after the other, each with a 52-byte payload. Unless the head node is transmitting data, the payload initially is empty. When a node receives a cell, certain criteria apply:

• If the cell has a payload (is not empty) and is addressed to that node, it copies the data out of the payload and retransmits the empty cell (or if the node has traffic to send, it can send it now).

- If the cell has a payload but is not addressed to the node, the node simply repeats it on the downstream link.

- If the cell is empty and the node has traffic to send, it might be able to put it into the payload of the cell to transmit it.

- If the cell is empty and the node has no traffic to send, it just retransmits it.

This scheme eliminates the dead time spent waiting for a token in FDDI. But what would happen when a node has a long message to transmit? It would monopolize the bus, locking out all of the nodes between it and the destination node. This is where the "distributed queue" comes in. The idea is to use a queue for accessing the bus, so all nodes get their fair share. It is not possible to have a single "control" node maintain the queue, because messages would have to flow back and forth between that node and all of the others, resulting in unacceptable overhead for message passing.

DQDB avoids that problem by distributing the queue among the nodes. The counter rotation of cells on the buses plays a role here as well. Assume that Node 5 needs to send a message on Bus A to Node 4. An algorithm is used to implement the distribution of the queue:

1. Node 3 puts a "reservation" on Bus B by setting a bit in the header of a cell. Since the cells are small, it will not have to wait long until it relays a cell with the reservation bit available.

2. Because of counter rotation, the reservation flows "upstream" with respect to Bus A, so every upstream node sees it. Downstream nodes do not need to see it.

3. Meanwhile, Node 3 counts the number of reservations it sees coming along on Bus B from downstream nodes of bus A, adding one to its own "queue counter" for each one it sees.

4. Then Node 3 gets an empty cell from Node 2. It looks at its queue counter. If the counter is greater then zero, it passes the empty cell along and subtracts one from the counter. Each time it does that, it has provided the cell required to satisfy one reservation from a downstream node. It does not know which node, but that doesn't matter.

5. Eventually, an empty cell comes along when Node 3's queue counter is zero. It knows then that no reservations are outstanding from downstream nodes, so it can go ahead and use the empty cell to send its message. If its message is longer than one cell, it must repeat the process until the entire message has been sent.

The opposite process is used to send messages on Bus B.

Switched Multimegabit Data Service (SMDS)

SMDS was created as a Metropolitan Area Network (MAN) service by Bellcore. In the purest sense, SMDS is a service definition and not a protocol. Its first realization was defined using the Distributed Queue Dual Bus (DQDB) technology, as specified in the IEEE 802.6 standard. The IEEE 802.6 DQDB standard defines a connectionless data transport service using 53-byte slots to provide integrated data, video, and voice services over a Metropolitan Area Network which is typically a geographic area of diameter less than 150 km. (90 mi.). The SMDS implementations based upon the IEEE 802.6 standard were the first public services to use ATM-like technology. Although the IEEE 802.6 standard also defines connection-oriented isochronous services, SMDS supports only a connectionless datagram service primarily targeted for LAN interconnection. Its name says a lot about SMDS. It is a MAN service (rather than a protocol). The fact that SMDS is switched means that to the networks that connect to it, it "has no distance." In other words, like an X.25 public network, all that is required to send a packet to another network connected to SMDS is the user's address. The subscriber does not see the nodes that are internal to SMDS.

Regarding the "multimegabit" in the name, SMDS falls in the same performance range as Frame Relay—T1/E1 to T3/E3—potentially overlapping the low end of B-ISDN

SMDS protocols are based on IEEE 802.6. However, SMDS supports only connectionless data traffic. The subscriber is attached to the SMDS service via T1 or T3. The Subscriber-Network Interface (SNI) defines the subscriber's interface to SMDS. To the subscriber's LAN, SMDS looks like another subnet. All of the internal nodes of SMDS are hidden from the subscriber, as with X.25. The interface with SMDS is the LLC layer. While using cell relay internally, SMDS is technically a frame switching facility, because LLC deals with frames. However, you are not likely to hear it called that.

The internal SMDS protocols are called SMDS Interface Protocol-1 , -2, and -3 (SIP-1 through -3). They are a subset of IEEE 802.6 (See Figure 11-34). Their correspondence is shown in the illustration. SMDS uses cell relay at layer 1, employing a cell with the same format as described above for ATM; however, the format of the 5-byte header is different. The SIP protocols were defined for use by the phone companies.

Several of the Bell operating companies have made SMDS available within their areas, and subscribers are beginning to test applications of SMDS/IEEE 802.6

SMDS and the IEEE 802.6 DQDB protocol have a one-to-one mapping. The SMDS Interface Protocol (SIP) has Protocol Data Units (PDUs) at layers 2 and 3. The layer 2 SIP PDU corresponds to the Distributed Queue Dual Bus (DQDB) Media Access Control (MAC) PDU of the IEEE 802.6 standard. The layer 3 SIP PDU is treated as the upper layers in IEEE 802.6. There is also a strong correspondence between these layers and the OSI reference model as shown.

Internationally a close relative of SMDS is the Connectionless Broadband Data Service (CBDS) as defined by the European Telecommunications Standards Institute (ETSI).

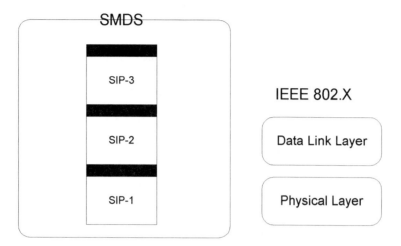

Figure 11-34. SMDS Protocols

Fibre Channel (Also *Fiber* Channel)

In 1988, ANSI chartered a working group with the task of developing a standard which would address the need for high data transfer rates of large amounts of data between workstations, mainframes, supercomputers, storage, and display devices. This working group was called the Fibre Channel working group. The Fibre Channel working group was charted with developing a set of standards with the primary objectives being:

- To provide high speed transfer of data between various types of end devices
- To provide a common transport mechanism for multiple higher layer protocols
- To support multiple physical layer interfaces
- To provide performance characteristics such as 1 gigabit/sec at distances up to 10km
- High bandwidth utilization of physical media

Networks and Channels

To provide the characteristics as stated in the overview, the solution would require a combination of features found in current networking and channel technologies. Networks and channels are the two basic types of connections that exist between processors in a communication environment (See Figure 11-35).

Channels are usually found in a highly structured environment (such as SNA) where the host and peripheral device configurations are known in advance. Such structure allows for channel connected communications to be highly efficient (error free) and able to achieve high levels of performance over short distances. Environments with high speed channels are master/slave arrangements where the host is the master and the devices attached, either local or remote, are slaves. Channels are the communication paths used to transfer the data between the host and peripheral devices. Channels are typically hardware intensive with little software processing overhead needed to accomplish data transfers.

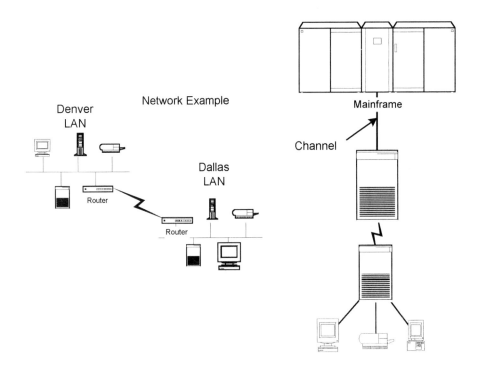

Figure 11-35. Networks and Channels

In contrast, networks operate in an unpredictable and less structured environment than channels. A network is a group of workstations, servers, and other peripheral devices which all communicate with a common set of protocols. This common set of protocols allows the network devices to communicate in a peer-to-peer fashion as opposed to the master/slave arrangement. The devices and hosts in a network operate independently and handle a wide range of tasks such as error detection and recovery and session management.

The Fibre Channel standard was initiated to meet the needs of both channel and network users. With Fibre Channel, both network users and channel users can share the same physical medium.

Information can be sent directly to a host, to a network (LAN to a workstation), or to other peripheral devices (See Figure 11-36).

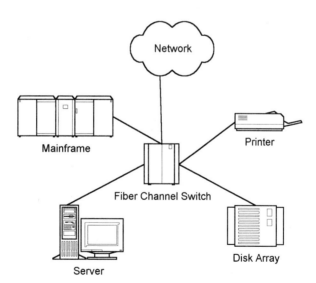

Figure 11-36. Fibre Channel Switch

Fibre Channel Summary

Fibre Channel is a high performance communication standard which will meet the needs of today and tomorrow's applications by providing a high-performance communications pathway as well as multiple vendor protocol support. It combines attributes of a channel as well as the attributes of a network for increased flexibility and usage.

Even though Fibre Channel is not widely implemented at this time, several efforts are underway to build products that utilize the standard. Most notably, the Fibre Channel Systems Initiative (FCSI) was created by IBM, Sun, and HP to advance the Fibre Channel standards process and to promote the use of Fibre Channel for high-performance workstations and devices.

SUMMARY

Today's digital telecommunications networks evolved out of the analog switched long distance networks that were created in the early part of the twentieth century. Digital telephony was developed for voice networks. It was only recently adapted for digital data networks. Prior to that, data communications required the conversion of digital signals to analog and back.

Analog voice signals are pulsating DC signals that vary from 0 to -48 volts and have a frequency range of 600 to 3400 Hz. Voice signals are transmitted between the subscriber and the Central Office (also local exchange or end office) on a two-wire local loop. An enterprise can install a Private Branch Exchange (PBX) to switch calls within the enterprise. PBXs are connected to local offices, and local offices are interconnected with trunks. A five-layer hierarchy has been defined for switching, with the local office at the lowest level and regional centers at the highest level.

Two technologies are important for analog voice networks: duplexing and Frequency-Division Multiplexing (FDM). Duplexed circuits have two wire-pairs rather than one, to allow half-duplex (nonsimultaneous transmission in opposite directions) and full-duplex (simultaneous transmission in opposite directions), and to simplify the amplification of signals for transmission over long distances. A FDM group places twelve voice conversations on one wire-pair by mixing twelve carrier frequencies (which differ by 4000 Hz) with the voice channels, then modulating the output with all twelve signals. Groups can be further multiplexed to form a supergroup and so on, up to a maximum of 900 conversations on an analog link.

In the 1960s, digital switches began to replace the previous electromechanical rotary or cross-bar switches. Electronic switches are controlled by digital control signals. Initially, control signaling was placed in the portion of the 4000 Hz voice channel that was not used for voice. The latest switching technology, such as AT&T's Signaling System Number 7 (SS-7), uses a separate channel for Common Channel Signaling (CCS).

In the 1980s, all of the long distance lines in the U.S. were converted from analog to digital transmission. The local loop is connected to a coder / decoder (codec). The codec encodes voice signals by sampling the signal 8000 times per second, converting it into a string of binary numbers. A voice channel requires 64,000 bps to reproduce the human speaking voice accurately. Digital signals preserve the quality of a transmission because it is much easier for digital transmission equipment to distinguish digital information from noise.

Time Division Multiplexing (TDM) replaced FDM. TDM multiplexers interleave the output of twelve codecs into a multiplex frame. Three standards prevail for multiplexing, North American, European, and Japanese. All three are based on a DS0 channel that is the Pulse Code Modulation (PCM) output of a codec, and define five or six levels of successive multiplexing. They differ in the number of voice channels multiplexed onto a given level.

A digital channel bank is a device that combines many codecs, a TDM, and a line driver to transmit framed output from twenty-four to ninety-six voice channels on a T-Carrier (T1, T1C or T1D) link.

A multiplexer combines the output of channel banks or other multiplexers onto a high-speed line. For example, a M1-3 combines the output of D1 channel banks onto a T2 link.

Digital signals are transmitted by varying the voltage on the link between zero and, usually, a positive and a negative value. Most often, it is the transition between levels, rather than the levels themselves, that signals a "1" or a "0" bit. Binary Eight Zero Suppression (B8ZS) is one of the most common encoding formats. It ensures that at least a one-level transition takes place for each byte transmitted.

The equipment that operates at OSI Layer 2 and above in the subscriber's office (usually a computer or terminal) is called Data Terminal Equipment (DTE). It connects to an analog phone line through Data Circuit-Terminating Equipment (DCE) and uses the RS-232 family of protocols. Modulators-demodulators (modems) operate in pairs at either end of a phone line to transmit bits through a variety of protocols. DTEs connect to digital communications links through Data Service Units (DSU) that connect through Channel Service Units (CSUs) which connect to the digital link. Multiplexers and Automated PBXs (PABXs) contain DSUs and CSUs.

Subrate digital data facilities are those that operate at less than the base voice channel (DS0) rate of 64 Kbps. T-carrier facilities include T1 and T3 (E1 and E3 in Europe). Fractional T1 (FT1) provides service between DS0 and T1 rates. Subscribers use T-carrier multiplexers to combine subrate channels into DS0 channels and to combine DS0 and larger channels into T1 or T3 channels. End Unit Multiplexers do not switch between T-carrier links, but Nodal Processor TDMs do, allowing the construction of complex networks with alternate paths.

X.25 currently is the dominant standard for packet switched networks. X.25 is a three-layer protocol (OSI Layers 1 through 3) and therefore provides error free, acknowledged service, with permanent virtual circuits (the equivalent of leased lines) and virtual calls (the equivalent of dialed lines) for the transmission of packets. Packet Assemblers/Dissassemblers (PAD) allow non-X.25 DTEs (such as dumb terminals) to send data over X.25 networks. X.25 provides DS0 throughput at best and has considerable overhead due to the acknowledged service it provides. X.25 will be supplanted by frame or cell relay technologies in the next few years.

Integrated Services Digital Network (ISDN) represents the first digital switched network that is not oriented toward voice communications. Instead, it is designed to serve multiple purposes: voice, data, video, facsimile, and all other forms of electronic communication. ISDN is both a set of services and a bundle of protocols. Services include 16 Kbps D channels for control signaling and other uses that do not require high data rates; 64 Kbps B channels for voice, data, and slow-scan video; and H channels with data rates of up to 1920 Kbps for full motion video, data network backbones, etc. Basic rate service provides D channel and two Bs. Primary rate service provides B and H channels that add up to T1/E1. Permanent circuits (like leased lines), circuit switched service, and packet switched service are provided.

ISDN protocols split the architecture into two parallel parts, one for data and one for control signals. Each has its own set of protocols at layers 2 and 3.

Broadband ISDN (B-ISDN) will use a form of cell relay called Asynchronous Transfer Mode (ATM). ATM operates at Layer 1 with small cells, rather than the large frames used by T-carrier technology. B-ISDN provides T3 and SONET and additional speeds up to 622 Mbps.

Frame Relay is currently offered by many packet switching companies. It operates at Layer 2 rather than Layer 3. It is faster and has less overhead than X.25, and it provides T1 data rates rather than X.25's DS0 rate. It can provide connectionless or connection-oriented service for data only.

IEEE 802.6 is the latest member of the 802 suite and is meant to provide a standard for Metropolitan Area Networks (MANs). It uses a Distributed Queue Dual Bus (DQDB) architecture and a form of Cell Relay that will become ATM at some future date. It provides better recovery from link failures and higher data rates than FDDI. 802.6 can provide connection-oriented or connectionless service for data, voice, and video.

Switched Multimegabit Data Service (SMDS) was developed by Bellcore to provide a MAN service with an IEEE 802.6 interface. It is analogous to X.25, but much faster. It adapts DQDB and ATM for an internal protocol that works for switched service, providing only connectionless service for data.

Synchronous Optical Net (SONET) is strictly a Layer 1 service from a data communications architecture perspective. It provides a standard for signaling on optical fibers so that multiple vendors can supply the network builders, and so that different networks can interconnect without the need to demultiplex optical signals. SONET also includes a very sophisticated network management capability, with a seven-layer protocol for use by the network manager and by subscribers.

Public switched networks within metropolitan areas in the U.S. are the province of Local Exchange Carriers (LECs), most of whom were formerly parts of AT&T and are called Bell Operating Companies (BOCs). LECs operate one or more Local Access and Transport Areas (LATAs), and they have a monopoly on services within their LATAs. However, they are prohibited from providing services between LATAs, even their own. Inter-LATA service is the province of Interexchange Carriers (IECs), who are represented in a LATA by a Point of Presence (POP). A subscriber must connect with sites within the same LATA through the LEC or with private facilities such as microwave. Connection to sites in other LATAs can be through the LEC to a POP, or the LEC can be bypassed with a private facility to the POP.

Public packet switching networks are built upon leased lines from the LECs or from private facilities. They provide dial access in the subscriber's local area, or a leased line from the subscriber's premises to the local access point can be obtained from the LEC.

Tariffs for long distance links are determined by a base charge, a per mile charge, and charges for optional services. Tariffs for packet networks are determined by access charges and a volume usage charge. When both LECs and an IEC are used to create a connection, multiple tariffs apply. Tariffs vary widely by supplier and by geographical area.

ADDITIONAL INFORMATION ON THE CD-ROM

- Additional ATM information
- LAPD traces
- Analog to Digital Conversion Algorithm
- RS-232 Pin Descriptions
- Null Modem Pin-out
- High-Level Data Link Control Protocol (HDLC)
- Channel Banks
- Analog Signaling
- Nodal Processors
- DTE/DCE examples
- Statistical TDMs
- T1 Framing example
- T1 Extended Superframes
- Modem Signaling

12

SAMPLE NETWORK

INTRODUCTION TO SAMPLE NETWORK

This chapter gives an example of a typical network in use today. This network consists of various architectures all rolled into one heterogenous network.

Figure 12-1 shows this campus network which consists of three locations. Only the main building is shown and a partial view of another location, building Y, is shown. The purpose of this section is to analyze the network and some of the individual components which make up the network.

CONNECTIVITY

The computer community consists of approximately 1,000 personal computers (PCs). These computers include over 100 different types of IBM and IBM clone PCs, around 150 UNIX workstations, and 300 Macintosh computers. The majority of this equipment is connected to some type of network. Obviously all of the individual components are not shown in Figure 12-1. This is true of almost any networking diagram; each individual PC or workstation is not shown, only the network to which the machine is attached and the major points of connectivity. For instance, point M on the diagram is a Token Ring network connected to the backbone which consists of fifty PCs and twenty workstations, but only four PCs are shown.

Computer users on this network have access to three types of network topologies: Novell Local Area Networks, an Ethernet network that consists mainly of UNIX workstations, and an AppleTalk network. All of these networks are interconnected into one network so that data, peripherals, and electronic mail can be shared by users on the networks. This illustrates the concepts of internetworking and interoperability.

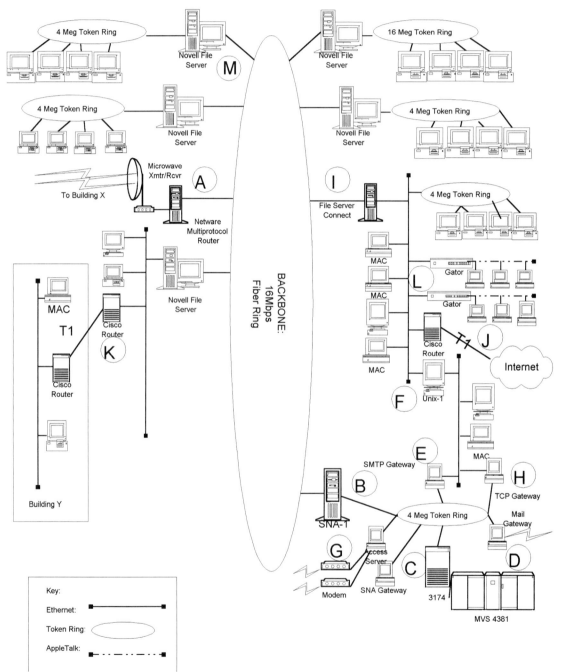

Figure 12-1. Network Example

Novell Local Area Network

There are Novell file servers with attached workstations as shown at point M. Two types of IBM PCs are utilized as Novell file servers in this particular network: the IBM model 80 (386/sx) PC and the IBM model 95 (4-86/sx) PC. The Token Rings are 4- Mbps Token-Ring Novell LANs, and one 16 Mbps Token-Ring Novell LAN, and additional Ethernet Novell networks. Unshielded twisted pair cabling is used on all of the Novell networks with one exception. The 16 Mbps network uses shielded twisted pair cabling. Novell NetWare version 3.12 is running on all of the Novell file servers.

Novell Network Interface Cards

There are no standards for workstations at this corporation. As a result, all makes and models of personal computers are connected to the Novell LANs. These range from XT models to 486 and Pentium based PCs. Connected workstations also include desktop PCs and laptops. Because networking capabilities are not part of the operating systems on PCs, all of the workstations that are connected to the Novell file servers contain a network interface card (NIC). Six kinds of NICs are used in these PCs. The PCs on the 4 Mbps Token-Ring networks use either an IBM Token-Ring Network PC adapter card or a 3Com Token-Ring adapter board. Laptop models use Xircom adapters connected to the parallel port. Although IBM was the card of choice when the first Token-Ring networks were established, 3Com cards were used later because they were less expensive. However, some problems have been encountered with PCs using the 3Com cards with recent releases of DOS, so the IBM 16/4 Mbps cards have become the standard NIC for Token-Ring PCs at this corporation.

    The 3Com Token-Ring NICs contain an RJ 11 telephone connector that enables the PC to be connected to a wall outlet. RJ 11 connectors are not part of the IBM card. You must attach a media filter to the IBM card when you are using unshielded, twisted pair cabling. The filter contains the RJ 11 phone connector. When a PC is on the Token-Ring 16 Mbps LAN that uses shielded twisted pair, a hermaphroditic adapter is used.

    A twisted-pair cable connects the PC from the RJ connector to an outlet, on the office wall. Wiring from the wall outlet is connected to a punch down block and from there, the wires are connected to a Multistation Access Unit (MAU). This Star Token-Ring topology configuration is the basic configuration. The MAU connects to a Token-Ring card in the file server.

    The PCs on the Novell Ethernet networks use 3Com NICs. They include 3C503, 3C507, and 3C509. Laptops on these networks use the Xircom Ethernet adapter which plugs into the parallel port of the PC. All of the Ethernet NICs contain an RJ45 connector. Like the Token-Ring PCs, a twisted pair cable connects the PC from the NIC to a wall outlet. From the outlet the cable is connected to a punchdown block. From the punchdown block they go to MAUs which are connected to additional MAUs to form the complete ring network (See Figure 12-2).

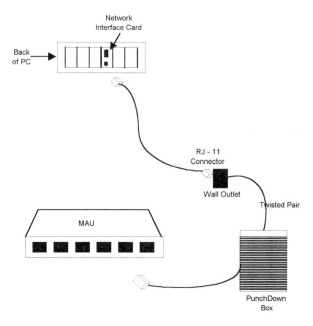

Figure 12-2. PC and MAU connectivity

Network Backbone

The Novell LANs are interconnected by a 16 Mbps fiber optic cable that is the backbone of the network. This media was chosen because it allows faster transmission speeds and because it can extend further distances than copper cabling.

Each of the Novell file servers is connected to the backbone by an Optical Data 836 Token-Ring transceiver. One end of the transceiver is plugged into a card in the file server using twisted pair cabling. Optic fiber cable comes out of the other end of the transceiver and it is connected to an Optical Data System 841 Token-Ring 16/4-Mbps MAU.

Main points of connectivity off the Backbone

Multiprotocol Router

A PC with a NetWare Multiprotocol Router, (circle A on Figure 12-1) is a device that also exists on the backbone. This router provides connectivity to another Novell, 4 Mbps, Token-Ring LAN that is 1.5 miles from the backbone to a site known as Building X. It routes data over the microwave to a Novell file server at this site.

SNA-1 File Server (SNA Connectivity)

The Novell Token-Ring file server called SNA-1 (circle B on Figure 12-1), provides several points of connectivity. This file server contains both a Token-Ring card and an Ethernet card. The file servers are connected to a MAU. On this ring is a dedicated PC that contains the SNA Gateway (circle C on Figure 12-1). This Gateway provides connectivity to an IBM 4381 host that contains some of the corporations mission-critical financial programs. All users on the Novell network can access this mainframe if they have established an account. The SNA Gateway converts the IPX packets to SNA format and sends the data to the IBM 3174 controller. The 3174 controller will allow 128 simultaneous sessions on the mainframe host.

Electronic Mail Connectivity

The primary cc:Mail PO (Post Office) also resides on this file server. Every Novell file server has a local PO that distributes mail to the users on that file server. When a piece of mail is destined for a user who is not part of the local PO, it is routed to the PO on the SNA-1 file server which will route it to the correct local PO. People who are not physically connected to the LANs are able to send and receive cc:Mail through the PO on the file server. This is done by two methods. The first method involves stand-alone users. Since cc:Mail is a proprietary mail package, users not connected to the LANs must purchase a copy of remote cc:Mail. An ID on the PO is established for the remote user by the cc:Mail administrator. Once this bit of housekeeping is done the user can dial into one of three cc:Mail Gateways (circle D on Figure 12-1 shows one of these). The Gateways are connected to the PO and mail is then distributed to the correct party by the PO.

cc:Mail POs from outside can also dial into the PO over the Gateways. However, the cc:Mail administrators on each of these independent POs must set up the hardware and software to enable this to happen. The cc:Mail system is operational seven days a week with a brief downtime of one hour. Steps are being taken to enable the PO to be operational at all times.

cc:Mail users can also send and receive mail to the Internet. Another piece of equipment on this LAN (circle E on Figure 12-1), the SMTP Gateway, enables this to occur. The SMTP Gateway is a dedicated PC that contains a gateway card. One part of the SMTP gateway is connected to the SNA-1 Novell file server and another part of it is connected to the UNIX workstation UNIX-1 (circle F on Figure 12-1). Using the IP address of the UNIX-1 Server, mail is directed to the Internet world on an Ethernet connection located on the other side of the UNIX-1 server. Translation of the packets occurs within the SMTP Gateway.

To users of cc:Mail, this is a seamless operation. To address mail to Internet users, Novell users select the name SMTP from their cc:Mail directory. They are then prompted to type in the Internet address of the person that they are corresponding with. Users can also transmit documents that were not created using the cc:Mail software. These documents include word processing documents, spreadsheets, and graphic files.

Dial-In Access to Novell LANs

Users can also access the Novell networks by using the Access Server, (circle G on Figure 12-1). The access server is a dedicated PC with FlexCom software that allows users to dial into the Novell Network. The user needs to have a PC, a copy of FlexCom Remote software, an ID on the Novell network, and a piece of equipment known as a secure key. The secure key is loaded with a group of numbers that were generated by the software on the Access server. A database on the Access server keeps track of the numbers that were generated and which user was assigned a particular set of numbers. When the user wants to log into the Access Server, the secure key displays one of these numbers. The user then keys in this number. If the number on the Access server matches the keyed number, then the user is permitted to enter the system. Up to seven users can access the Novell networks in this manner at any one time. There are twenty users currently accessing the Novell networks in this way and contention is minimal.

TCP Gateway

Users from the Novell networks can access workstations on the UNIX network by using either Novell's LAN WorkPlace for DOS product or Wollongong's WIN/TCP product. The Wollongong TCP Gateway, (circle H on Figure 12-1), is connected to the SNA-1 file server and it is also connected to the UNIX workstation, UNIX-1, circle F. Users on the Novell networks must have a Novell interface card and the WIN/TCP software installed on their PCs. This software enables Novell users to access all the TCP/IP services that exist on the UNIX network.

File Server "Connect"

A Token-Ring file server on the main campus, (circle I on Figure 12-1), supports a 4 Mbps Token-Ring network of about fifty PCs. It is also the main point of connectivity between the Novell network and the UNIX network because it also contains a 3Com 532 card which provides Novell connectivity to the Ethernet world.

Ethernet Internal Connectivity

The Ethernet network is divided into three segments, with two of the segments coming into the main campus from Building X and Building Y. The UNIX workstation, UNIX-1, contains two controller cards that enable it to connect to two of these segments. One of these segments is the same segment that File Server "Connect" is connected to. Packets that come from the Novell backbone that are destined for stations on the Ethernet network are routed through this workstation.

The types of equipment that are on the main Ethernet segment include various types of SUN workstations, RISC 6000 machines, Macintosh computers with Ethernet cards, Gator boxes, and three Cisco routers.

Ethernet Connectivity to Building Y

The Cisco router, (circle K on Figure 12-1), provides connectivity to another building in this campus network, Building Y. This building is twelve miles from the main campus and is connected via a T1 link. The T1 line connects two Cisco routers which are located at each site. Data is routed from the Cisco router to the Novell Ethernet network, UNIX workstations, or Macintosh machines that contain Ethernet cards.

AppleTalk Connectivity

The main campus contains two AppleTalk networks (circle L on Figure 12-1). The AppleTalk network uses the extra pair from the telephone cable to physically connect the Macintosh workstations. These networks are connected to the UNIX world by one of three Gator boxes. These Gator boxes are gateways that translate the AppleTalk protocol to Ethernet protocols. The Macintosh computers can access all of the UNIX world, including the Internet.

Ethernet Connectivity to the Outside World

The second Cisco router (circle J on Figure 12-1) also has a T1 line connected to it. Information passing over this line is destined for a node connected to the Internet access provider Colorado Supernet. This is the primary connecting point to the Internet. All users on the network who have TCP/IP software are able to access any site that is connected to the Internet. Since TCP/IP is built into the operating system of the UNIX and MAC boxes, it is easy for them to access other computers on the Internet that are outside of the campus network.

LAN WorkPlace For DOS

PC users on the Novell Network can access the TCP/IP protocol suite of services by having Novell's product, LAN WorkPlace for DOS, installed on their PC. This is a product that can be used from either the DOS prompt or from the Windows environment. The PCs must have at least 425K of available RAM, 2.8 MB of free hard disk space, and a Novell Ethernet card that supports Novell's Open DataLink Interface (ODI). The LWP/DOS software must be installed on each computer. The ODI files perform the traditional network access layer functions for the

LWP/DOS software's implementation of TCP/IP. ODI adds functionality to NetWare and network computing environments by allowing multiple protocols to bind to, or operate with multiple network drivers simultaneously. ODI creates a logical network board to send different packet types over one network adapter card and wire.

SUMMARY

This example represents a networking environment from a medium size corporation in use today. As the corporation grows, the network will continue to grow as well. Connectivity solutions will continue to be explored which will allow users on the network to access noncompatible components and systems. The backbone of the network will be updated soon to give the user community better performance when utilizing resources across the network.

ADDITIONAL INFORMATION FOUND ON THE CD-ROM

- Additional Network Diagrams with descriptions
- Explanations of concepts used in this chapter

GLOSSARY

NOTE: A much more extensive glossary is found on the CD-ROM. Use the search function to access over 3000 terms and definitions located on the CD-ROM, or go directly to the glossary from the table of contents.

10Base2

IEEE standard for 10Mbps baseband Ethernet over coaxial cable (RG-58). 185 meters is the maximum distance for this standard. 10Base2 is also referred to as "THINNET," "THINLAN," and "CHEAPERNET." 10Base2 or Thinnet uses twist-on BNC connectors to attach to devices. Each device that attaches uses a T-connector to daisy-chain to the next device. The final T-connector in the series must include a termination plug. In most 10Base2 implementations the Network Interface Card contains the transceiver functions.

10Base5

IEEE standard for 10Mbps baseband Ethernet over coaxial cable. 500 meters is the maximum distance for this type of cable. Also called "Thicknet" and "Yellow Wire." This type of physical cable is typically used as the backbone media for Ethernet networks.

10Base T

Standard for Ethernet media using twisted pair similar to modular telephone cabling. 10Base T networks use twisted pair between the workstation and a hub. The hub is then attached to the backbone of the network. This arrangement isolates each workstation from the backbone. The segment that extends from the workstation to the hub is often referred to as the "home run."

2B+D

This is the common way of stating ISDN's BRI interface. 2B stands for two bearer channels. D stands for a data channel. BRI is the Basic Rate Interface in ISDN. On ISDN circuit is composed of two 64Kbps bearer channels which may be used for voice or data, and one 16Kbps data channel. The data channel is also used for signaling.

23B+D

The ISDN Primary Rate Interface. It is a circuit consisting of 23 64Kbps channels for carrying voice, data, and video and a data channel used for carrying signaling information. It is similar to T1 signaling.

3270

A type of IBM terminal or printer used in SNA networks. Other vendors also provide 3270 emulations for their terminals and printers.

3270 Gateway

A computer system which handles the communication path and conversion between a terminal device or PC and an IBM mainframe.

3274

IBM cluster controllers or communications controllers. These devices are used to control communications between an IBM mainframe and a terminal device. The devices can be 3270s or ASCII terminals.

3745

IBM Front End Processor (FEP) model number. Older versions included the 3725 and the 3705. These devices attach LANs and other devices such as cluster controllers to an IBM host. 3745s can also be attached to each other in cross domain configurations.

802.1

Standard for overall architecture of LANs and internetworking.

802.1B

Standard for network management.

802.1D

MAC layer standard for inter-LAN bridges. This standard encompasses the interconnection of 802.3, 802.4, and 802.5 LANs.

802.2

Standard for Upper Data Link Layer sublayer also known as the bridge (Logical Link Control) layer. It is used with the 802.3, 802.4, and 802.5 standards (lower DL sublayers).

802.3

Standard for CSMA/CD (Carrier Sense Multiple Access with Collision Detection). Both Ethernet and Starlan follow this standard. It encompasses both the MAC and physical layer standards. At the Data Link layer, it is one of three primary data link sublayers. The physical layer specifications depend on the type of media used (10Base T, 10Base 5, etc.). 10Mbps is the transmission rate for this standard.

802.4

Standard for the Data Link and Physical Layer for the Token Bus protocol. It is typically used with the Manufacturing Automation Protocol (MAP) developed by General Motors. 10Mbps is the typical transmission speed for this standard. See the LAN Architectures section of this document for further details on this subject.

802.5

Standard for the token-passing (Token Ring) access method of LAN protocols. It encompasses both the Data Link and Physical Layer standards. Transmission speeds include both 16Kbps and 4Kbps. See the LAN Architectures section of this document for further details on this subject.

802.6

Standard for Metropolitan Area Networks (MANs) also known as Distributed Queue Dual Bus (DQDB). For further details on this subject see Telecommunications, Cell Relay, B-ISDN, and ATM tutorials.

Adapter Card

An adpater card is a printed circuit card which fits inside a computer and communicates information from the computer to another computer or printer. It takes data from the computer's memory and transfers the data via a coax cable, RS232, cable or similar connection. An example would be a 3278 adapter card which fits inside a PC and emulates a 3278 terminal. Network Interface Cards are examples of specialized adapter cards.

Adaptive Routing

Adaptive routing is a method of routing packets (datagrams or messages) in which the computers along the communications path select the best route for the packet to take. When network congestion or link failures occur, the path the packet takes changes or adapts.

Address

An address is a unique identifier which determines the originating location of data or the destination of data being transmitted across a communications link. There is a distinction between a station's link address, a station's network address, and an individual process running on a station.

Address Mask

An address mask is a bit mask used to select bits from an Internet address for subnet addressing. The mask is 32 bits long and selects the network portion of the Internet address and one or more bits of the local portion. It is also called a subnet mask.

Address Resolution Protocol (ARP)

ARP is the Data Link Layer protocol used by IP (as in TCP/IP) for address resolution. Address resolution refers to the ability of a machine to resolve another station's MAC (hardware) address given its IP address.

Anonymous FTP

Anonymous FTP allows users who do not have a password or login ID to access certain files on remote machines. Users can get public domain software from Internet sites using "anonymous" as the user ID and their userid@hostname.domain as their password. A database called "ARCHIE" details the software that is available using the anonymous FTP method.

ANSI

American National Standards Institute is a national voluntary organization which develops and publishes standards. It develops standards for data communications, programming languages, magnetic storage media, the OSI model, office systems, and encryption. It is a member of the International Organization for Standardization (ISO). For the most part, ANSI standards are similar to ISO standards, differing only when unique aspects of North American systems need to be considered.

API

API stands for Application Programming Interface. An API is a formally defined programming language interface. It is software that allows application programs to interface to lower level services performed by a computer's operating system. Application programs make function calls which perform these lower level functions such as transferring files. Programmers can use these function calls to gain access to services provided by an operating system. NetBIOS can serve as an API for data exchange.

APPC

APPC stands for Advanced Peer-to-Peer Communications. It is an IBM protocol allowing IBM network nodes to communicate in a "peer" arrangement as opposed to the typical SNA hierarchical arrangement. APPC consists essentially of two "new" Network Addressable Unit, LU 6.2. and PU 2.1.

AppleShare

Appleshare is Apple Computer's Local Area Network where nodes use the AppleTalk protocols to communicate. It allows sharing of files and network services between Apple computers using the AppleTalk, EtherTalk, and TokenTalk topologies.

AppleTalk

AppleTalk is Apple Computer, Inc.'s LAN for connecting Macintosh computers, printers, and other resources together.

APPN

APPN stands for Advanced Peer-to-Peer Networking. It is an SNA feature offered by IBM which allows more efficient routing of data between nodes in an SNA network. It allows remote workstations to communicate in a peer-to-peer fashion as opposed to a hierarchical arrangement where communication is controlled via a host computer. It was introduced by IBM in 1985 as an alternative to the growing popularity of client/server computing technologies.

ARCHIE

Archie is a computer program that can be located on the Internet at several different locations. Its function is to provide a listing of files available on the Internet and their locations.

ARCNET

ARCNET stands for Attached Resource Computer Network. It was one of the earliest Local Area Networks (developed by Datapoint Corp). It uses the Token Passing scheme at 2.5 Mbps in both star and bus topologies.

ARM

ARM stands for Asynchronous Response Mode. It is a mode of communication where either a primary or secondary station can initiate transmission of a message.

ARQ

ARQ stands for Automatic Retransmission Request. It also stands for Automatic Request for Repeat. In the event of an error, data communications protocols deal with retransmission of data using the ARQ method. Essentially when a receiving station receives a message (frame) containing corrupted data, it responds with an ARQ which requests the retransmission of the errored frame.

ASCII

ASCII, American Standards Committee for Information Interchange, is one of the two most widely used codes (EBCDIC is the other). Codes represent characters, such as keyboard characters. ASCII uses 7 bits for the 128 elements it represents. For example, when the character "A" is pressed on the keyboard, the ASCII binary representation is 100 0001 (hexidecimal 41).

ASN.1

ASN.1 stands for Abstract Syntax Notation One. It provides a standardized format for data transfer between nodes. Each node is only concerned with translating to and from ASN.1 and does not need to know anything about the format in which data are stored elsewhere on the network.

Asynchronous Transfer Mode

Asynchronous Transfer Mode, or ATM, is a cell relay technology based on small (53-byte) cells. An ATM network consists of multiple ATM switches which forward each individual cell to its final destination. ATM can provide transport services for audio, data, and video.

Asynchronous Transmission

Asynchronous operation means simply that bits are not transmitted on any strict timetable. The start of each character is indicated by transmitting a start bit. After the final bit of the character is transmitted, a stop bit is sent, indicating the end of the character. The modems must stay in synchronization only for the length of time that it takes to transmit the eight bits. If their clocks are slightly out of synch, data transfer will still be successful.

B-ISDN

B-ISDN stands for Broadband ISDN. ISDN line rates come in three basic varieties, basic, primary, and broadband. Basic or "Narrow" ISDN consists of two bearer (B) channels and a data (D) channel. Each bearer channel can carry one PCM voice conversation or data at a transmission rate of 64Kbps. The ISDN Primary Rate Interface consists of twenty-three 64Kbps channels for carrying voice, data, and video and a data channel used for carrying signaling information. It is similar to T1 signaling. B-ISDN (also called wide ISDN) has multiple channels above the primary rate. In addition to the B and D channels, there are a number of additional channels defined. These include the A, C, and H series of channels.

Backbone

The backbone of a network is the portion of the network which carries the most significant traffic. It is also the part of the network which connects many LANs or subnetworks together to form a network. Bridges are often used to form network backbones. In this configuration, bridges often limit the local traffic from the backbone to reduce congestion and isolate problems.

Backward Learning

Backward learning is a process or algorithm that takes place in a network computer used for routing packets or datagrams. As packets traverse a network and go through servers, the server will learn routes based on the information contained in the source and destination fields of the packets. The backward routing algorithm is the algorithm used to determine future routes based on this information.

Bandwidth

Bandwidth is the difference between the highest and lowest frequencies that can be transmitted across a transmission line or through a network. It is measured in Hertz (Hz) for analog networks and Bits Per Second (bps) for digital networks.

Barrel Connector

A barrel connector is a barrel shaped connector which joins two 10Base5 (thicknet) pieces of Ethernet coax together.

Baseband

Baseband is a form of modulation where signals are placed directly on the transmission media in their original form, i.e. no frequency division multiplexing is performed. A baseband modem is a modem which does not modulate the signal before transmission, thereby transmitting the signal in its native form. Baseband signaling is the transmission of either digital or analog signals at their original frequencies.

Basic Rate Interface

BRI is the Basic Rate Interface in ISDN. A BRI ISDN circuit is composed of two 64Kbps bearer channels which may be used for voice or data, and one 16Kbps data channel.

Baud Rate

Baud rate is synonymous with the signaling rate. Baud rate and Bit rate (bps) are often confused. The strict definition of baud rate refers to the number of times per second the signal on the communication link changes. In other words, if a modem is able to shift between frequencies 9600 times per second, then the baud rate is 9600. If one bit is represented by each frequency shift the baud rate and bit rate would be the same, 9600. However, if each frequency shift represented two bits, then the baud rate would be 9600 and the bit rate would be 9600 X 2 = 19,200 bps.

Beacon

A beacon is a type of Token Ring frame which is used to indicate a catastrophic condition on the Physical ring. Conditions such as broken ring cables or malfunctioning workstations or MAUs (Multistation Access Unit) will create this condition. Beaconing refers to the recovery process which takes place when such situations exist.

Binary

Binary is the numbering system used by computers to represent information. Binary numbers consist of numbers represented by two values, 1s and 0s. Most computers store information in bytes (and multiples of bytes) which consist of 8 bits. These 8 bits can be combined to form 256 different values. The values represented by these 8 bits would be between 0 and 255. 0 would be represented by 00000000, Decimal 1 by 00000001, Decimal 2 by 00000010, Decimal 3 by 00000011, Decimal 4 by 00000100, and so forth until 255 which is represented by 11111111.

Bindery

A Bindery is a Novell NetWare database which contains definitions for users, groups, and work-groups. It contains three components: objects, properties, and property data sets. Objects include both logical and physical entities such as users, servers, printers, and so forth. Properties are characteristics of each object and contain information such as addresses, rights, and pass-words. Property data sets are the values assigned to the properties.

BIOS

BIOS stands for Basic Input/Output System. The BIOS is a computer storage area (buffer) which stores data to be sent to a computer's hardware components such as a serial interface adapter.

Bis

Bis, as seen following some of the ITU-T standards, is French for second. Therefore, a standard such as V.22 bis means that it is the second V.22 standard defined. Ter is French for third and is also found after standards when it is the third time the standard has been defined.

Bisync

Bisync is short for Bisynchronous Transmission. Bisynchronous Transmission is a synchronous character-oriented protocol developed by IBM in 1964. It is still widely used even though the newer bit oriented protocols used today are much more efficient.

Bit Interleaving

Bit interleaving refers to the process of interleaving bits within a serial communications link. As depicted below, lower speed transmission links are time division multiplexed onto a higher speed communications link by interleaving bits from each channel onto the high speed link. It must be noted that depending on the technology, bits, or bytes may be interleaved. T1 transmission inter-leaves bytes from codecs.

Bit Rate

The bit rate is the number of bits transmitted per second (bps). For instance, commonly used rates for modem speeds are 300 bps, 1200 bps, 9600 bps, 14,400 bps, and 19,200 bps. Another way of writing 19,200 bps is 19.2Kbps. Kbps stands for Kilobits per second which would mean 1000 bits transmitted per second times 19.2. 1.544 Mbps is the T1 rate and is 1,544,000 bps.

Bridge

A bridge is a hardware device which connects LANs together. It can be used to connect LANs of the same type, such as two Token Ring segments, or LANs with different types of media such as Ethernet and Token Ring. It operates at the Data Link Layer of the OSI Reference Model as shown in the figure below. The diagram also shows a bridge connecting several Ethernet segments together to form one single network from several subnets. For an indepth description of bridges and bridge types, see the Bridges section of the Internetworking Tutorial.

Broadband

Two methods are used to transmit signals between nodes: baseband and broadband. In the context of LANs, broadband refers to analog transmission of digital signals and baseband refers to digital transmission of digital signals.

Broadcast

The term broadcast is used to indicate the process of sending information over a network to more than one device simultaneously. Several broadcast terms are used in the data communications industry.

Brouter

Also written as B/Router, this is a node that performs the functions of both a bridge and a router, for example, bridging frames between two subnets but routing packets from either subnet to another network. Its advantage is flexibility. For example, the brouter routes traffic from an Ethernet LAN to a Token Ring LAN and bridges traffic from an Ethernet LAN to another Ethernet LAN.

Bus

There are many types of buses. There are Internal Buses, External Buses, and LANs which operate on bus topologies. Internal buses are buses within a PC which connect the central processor with the video controller, disk controller, hard drives, and memory. External buses are buses such as AT, ISA, EISA, and MCA which are internal to a PC but not "Local Buses." Network buses refer to the type of topology that is used when connecting nodes in a Local Area Network environment. A network with a bus topology has a single wire or cable with multiple taps; the ends of the bus are unconnected.

Cache

Cache memory is memory that is readily accessible to a computer's central processor. Disk cache memory is a portion of RAM (random access memory) which is available for storage and retrieval of commonly used data. Using disk cache speeds up computer operations because accessing data stored in RAM is much faster than accessing data stored on a hard drive. When a process running in a computer needs access to data, cache memory is checked first. If the data is present, it is retrieved.

Capacity

Capacity is the capability of a data communications component to carry out its intended function. It is typically used to describe the capability of a communications channel or link. For instance, the capacity of a T1 channel is 64Kbps. This does not mean that the channel always contains 64Kbps of data, but that it is capable of carrying 64Kbps of data.

Carrier

A carrier is a company which provides communications circuits. Sometimes the digital communication channels, designated T1, T2, etc., are referred to as carriers as well. Note that various carriers are implemented with various transmission media (copper wire, coax cable, fiber optic, etc.) and employ varying digital signal encoding formats as appropriate for the medium.

Carrier Sense Multiple Access

CSMA/CD stands for Carrier Sense Multiple Access with Collision Detection. The inventors of Ethernet chose this technique for controlling access to the medium (the bus).

CDDI

CDDI stands for Copper Distributed Data Interface. It is a version of the Fiber Distributed Data Interface which runs on copper wiring such as twisted pair.

Cell Relay

Cell Relay is a packet switching technology that operates at the physical and data link layers of the OSI model. It uses a small, fixed-length cell. These small cells can carry any type of information such as data, voice, and video at tremendously high speeds. Each cell is composed of 53 bytes of which 5 bytes are overhead.

Central Office

A Central Office (CO) is a telephone company facility where local loops are terminated. The function of a CO is to connect individual telephones together through a series of switches. COs are tied together in a hierarchy for efficiency in switching. Other terms for a Central Office are Local Exchange, Wiring Center, and Public Exchange.

Channel

Generically speaking, a channel is a communications path between two or more communicating devices. Channels are also referred to as links, lines, circuits, and paths. In the mainframe environment, the channel is the path between the host computer and a controller device.

Channel Attached

Channel attached devices are devices which are attached directly to a host computer. This term is usually used in reference to an IBM mainframe. Channel attach and local attach are synonymous terms. Locally attached devices are devices that are attached directly to the bus and tag cables (or ESCON on newer hosts) coming out of the mainframe. Typical local attach devices are high speed disk (DASD), high speed printers, Front End Processors, and Cluster Controllers.

Channel Bank

The devices that perform the multiplexing for FDM transmission are called channel banks. Similar devices for multiplexing digitized voice signals are also called channel banks. Channel banks take many slow speed voice or data links and multiplex them onto a high speed channel such as a T1.

Channel Service Unit

A Channel Service Unit or CSU is a device that connects customer equipment to digital transmission facilities such as a T1 circuit. The primary functions of a CSU are line conditioning, equalization, and loopback (for testing purposes). The local loop for digital service always terminates at a Channel Service Unit (CSU) in the subscribers building. The CSU is the device that actually generates the transmission signals on the local loop (that is, the telephone channel).

Circuit

A circuit is the physical connection between two communicating devices. See also channel, circuit-switched networks, and virtual circuits.

Circuit Switching

Circuit-switched networks establish a physical connection between two nodes, and a packet is passed between nodes by "switching" it through intermediate points, either through other nodes or through a host computer.

Client Server

Client Server or Client/Server is a mode in computer networking where individual PCs can access data or services from a common high performance computer. For instance when a PC needs data from a common database located on a computer on a Local Area Network, the PC is the client and the network computer is the server.

CLNP

CLNP stands for Connectionless Network Protocol. It is an Open Layer 3 protocol (Network Layer).

CODEC

CODEC is a contraction which stands for coder decoder. It is a hardware device which takes an analog signal and converts it to a digital representation.

Collapsed Backbone

A network topology which consists of a hub as the central point of the network is often called a collapsed backbone. This is in contrast to standard backbones which, in the case of Ethernet, consist of a single common cable accessed by individual subnets.

Compression

The term compression refers to the process of reducing the number of bits required to represent data without altering the meaning of the information being conveyed by the data. The primary reason for using compression techniques is to optimize the use of the communications channel. It is a function of the presentation layer of the OSI model.

Conditioning

Conditioning refers to the process of modifying transmission facilities to reduce noise. This is typically done on leased lines for the purpose transmitting data, which is more susceptible to noise than voice signals.

Connection Oriented

The term connection oriented refers to a protocol mode used in data communications networks where the send and receive stations in a network remain connected for the duration of the session. A session is established between the sending and receiving stations for the duration of the session. The stations stay in contact during the session while the datagrams are being sent back and forth.

Connectionless

Connectionless Transmission refers to a protocol mode where each individual packet (or datagram) is an independent unit. The network is responsible for getting each datagram to the final destination. The network does not have to establish and maintain a session before sending and receiving datagrams. A network may be connectionless or connection-oriented depending on the particular protocol used.

Control Character

Control Characters are ASCII characters such as Carriage Return (CR) and Line Feed (LF). They are nonprintable characters used for control of the flow of data or the format of data. They are entered from keyboards by holding down the control key while simultaneously pressing another key. For instance holding down the control key and pressing M and J keys produces the carriage return and line feed control characters.

CPE

CPE stands for Customer Provided Equipment or Customer Premise Equipment. It refers to telephone equipment that resides at the customer site.

CRC

CRC stands for Cyclic Redundancy Check. It is the mathematical process used to check the accuracy of the data being transmitted across a network. When a block of data is about to be sent from one station to another, it performs a calculation and appends a value on the end of the block. The receiving station takes the data and the CRC character and performs the same calculation to check the accuracy of the data.

D Channel

In ISDN, the D Channel is the Data Channel. For the ISDN Basic Rate Interface, the D channel is 16Kbps. In the Primary Rate Interface, the D channel is 64Kbps.

Dark Fiber

Dark Fiber is fiber optic cable which does not have light traveling through it. Fiber is said to "go dark," which means light (or a signal composed of light) is no longer present.

DARPA

DARPA stands for Defense Advanced Research Projects Agency. It is the government agency which sponsored the ARPANET and the Internet.

DASD

DASD stands for Direct Access Storage Device is an online storage device for host computers.

Data Link Control

Data Link Control are the characters used to control the transfer of data frames (such as MAC frames) between two nodes. These control codes perform such functions as flow control and sequencing.

Data Link Layer

The Data Link Layer is the second Layer of the OSI model. It resides between the Physical Layer and the Network Layer. It is responsible for point-to-point transfer of Data Frames.

Datagram

A datagram is a packet which is transferred independently of all other packets. The term datagram is usually used in reference to an IP layer packet which uses connectionless delivery of packets (datagrams). In other words a session is not established before sending a datagram from one node to another.

DCE

DCE stands for Data Circuit-Terminating Equipment. In a network, there are two broad categories of devices, DCEs and DTEs. DTE stands for Data Terminal Equipment. The difference between the two is that DTE is the end device in a network and the DCE is the device or devices in the network that transmit and receive the DTE data.

DDS

DDS stands for Digital Data Service (also Dataphone Digital Service). It is a series of services provided by the telephone company that provide digital facilities for data communications. DDS comes in several speeds, 2.4Kbps, 4.8Kbps, 9.6Kbps, and 56Kbps.

Device Driver

A device driver is a program that controls devices attached to a computer such as a printer or a hard disk drive.

DNA

DNA stands for Digital Network Architecture. It is Digital Equipment Corporation's system architecture.

DQDB

DQDB stands for Distributed Queue Dual Bus. For a more complete description of DQDB, refer to the Telecommunications Tutorial. It is the IEEE 802.6 standard for Metropolitan Area Networks (MANs).

Dumb Terminal

A dumb terminal is a terminal which is solely dependent on a host computer for processing capabilities. Dumb Terminals typically do not have a processor, hard drive, or floppy drives, only a keyboard, monitor, and a method of communicating to a host (usually through some type of controller). Intelligent terminals, on the other hand, contain their own processor and storage devices as well as software programs.

Early Token Release

Early Token Release is a method of releasing the control token on a Token Ring LAN before the sending station has received the transmitted data frame. This is done on 16Kbps Token Rings.

EBCDIC

EBCDIC stands for Extended Binary Coded Decimal Interchange Code. It is the IBM standard for binary encoding of characters.

Enterprise Network

An Enterprise Network is the same thing as a Corporate Network. It means the entire network of a single corporation. The goal of enterprise networking is to integrate all of the diverse systems within a single organization.

ESCON

ESCON stands for Enterprise System Connections. It is IBM's latest channel technology that provides dynamic connection of devices to a host over fiber optic links of up to 60 kilometers.

EtherTalk

EtherTalk is Apple's protocol for Ethernet networks.

Fast Ethernet

Fast Ethernet is Ethernet that runs at 100 Mbps as opposed to the standard 10 Mbps. There are two primary standards that have been proposed: 100Base-T and 100VG-AnyLAN. 100Base-T is also referred to as the IEEE 802.3u standard. The 100VG-AnyLAN standard is in the hands of a new committee, the IEEE 802.12. Fast Ethernet and the 100Base-T standard are synonymous terms.

Fast Packet Switching

Fast Packet Switching is a new packet switching technology that operates at the physical layer and lower Data Link sublayer of the OSI protocol stack. Because it is located at these low layers (only a small amount of processing is needed) and because of small packet sizes, it operates at very high speeds. Performance is also increased because error checking and recovery are handled by the end nodes (and not the network).

FDM

FDM stands for Frequency Division Multiplexing. Frequency division multiplexing was developed to allow several voice signals to be transmitted (multiplexed) simultaneously over a single trunk.

FEP

FEP stands for Front End Processor. A FEP is used by a mainframe to handle the details of the network when sending and receiving data from the network. This keeps the host processor from having to handle the processing associated with routing data through the network.

Flow Control

Flow control is a method of controlling the amount of frames or messages that are sent between two computer systems. Practically every data communications protocol contains some form of flow control to keep the sending computer from sending too many frames or packets to the receiving node.

Fractional T1

Fractional T1 (FT1) is a service offered by the phone company which provides users of telecommunications services optional data rates from 64Kbps to 1.544Mbps. It is called Fractional T1 because the user can specify the desired rate, which is a fraction of the normal T1 rate (1.544Mbps). It is low cost alternative to purchasing a full T1 and only using a portion of the bandwidth.

Fragmentation

The term fragmentation is usually used in the context of the Transmission Control Protocol/Internet Protocol (TCP/IP). It is the process that is used by the IP layer to divide large messages into smaller pieces before sending them to the Data Link Layer for transmission. It is necessary to do this because a protocol such as Ethernet has a maximum frame size of 1500 bytes. If the IP packet is larger than 1500 bytes, then fragmentation is used for the transfer.

Frame

A frame is a unit of information transmitted across a data link. There are two types of frames, a control frame and an information frame. Control frames are used for initial setup and management of the link. Examples of control frames (from the LAP-B protocol) are Receiver Ready (RR) and Receiver Not Ready (RNR) which tell the sender whether or not the receiver is capable of accepting frames. Information frames contain information from the layer above, the network layer.

Frame Relay

A frame relay is essentially an electronic switch. Physically, it is a box which connects to three or more high-speed links and routes data traffic between them. Frame relay is intended only for data communications, not voice or video. Only connection-oriented service is provided. Transmission errors are detected but not corrected (the frame is discarded).

Frequency Modulation

Frequency Modulation is a method of modifying a signal so that it can carry information. The carrier (original sine wave) has its frequency modified to correspond to the information being carried.

FTP

FTP stands for File Transfer Protocol. FTP is an application layer protocol used in TCP/IP networks. FTP is a program that can be used to transfer files between hosts on the Internet. There are two ways to use FTP: regular FTP and anonymous FTP. To use regular FTP you must have an account with the remote host. Anonymous FTP does not require an account and allows access to thousands of archives such as public domain software.

Full Duplex

Full-duplex means that data are transmitted in both directions at the same time, using one wire-pair for each direction. In principle, full duplex allows data to flow in both directions simultaneously.

Gateway

A gateway is a device which routes datagrams and/or packets from one network to another. The networks are typically different types so that a protocol conversion must take place as well.

Gopher

Gopher is an Internet program developed at the University of Minnesota which allows Internet users to find different resources and applications using a menu-driven system. It was designed primarily to act as a distributed document delivery system. These documents may reside on local or remote Internet hosts. The Gopher menu system abstracts the detail and makes it appear that everything is local.

H-Channel

The H-Channel is the channel on an ISDN Basic Rate Interface that is used for higher bit rates. The bit rates associated with these channels are H0 at 384kbps, H11 at 1536kbps and H12 at 1920kbps. These channels could be used for high bit rate communications such as video telephoning.

Half-Bridge

As an enterprise network grows, it tends to disperse over a wide geographical area. Another important use of bridges is to connect LANs remotely. WAN bridges, also called half-bridges, work together in pairs.

Half-Duplex

A leased line typically has four wires (although it is possible to lease a two-wire line). Half-duplex means that one wire-pair is used to transmit in one direction and the other in the other direction, but transmission does not take place in both directions at the same time. Half-duplex is faster than using a single wire-pair for both directions because it is not necessary for the modems to wait for the line to "turn around" each time the direction of data transmission reverses.

Handshaking

Handshaking refers to the initialization process that two or more computers go through before they are able to communicate. It is the first part of each and every data communications protocol. It is used to establish initial setup parameters.

Header

A header is part of a message, packet, or frame which contains information necessary to send a unit of information from one node to another. The header normally contains a field specifying the length of the encapsulated message together with at least one field providing information about the message. If, for example, the message is a segment of a larger message, the header might specify the relative position of the segment in the complete message and probably the total number of segments in the message.

Hexidecimal

The Hexidecimal Number system is a base 16 numbering system. It is the numbering system used to condense binary bytes into a compact form for printing or analysis of computer data. It is composed of the numbers 0-9 and the letters A-F. Each "nibble" (four bits) of a byte can be represented by one of the sixteen characters.

Host Computer

Generically speaking, it is a computer system that contains (or hosts) the necessary equipment software and hardware so that it can connect and operate in a network environment. In the IBM world, a host is a mainframe or Central Processing Unit. An example of an IBM host would be the ESCON 9121.

HPPI

HPPI stands for High Performance Parallel Interface. It is also referred to as HiPPI. HPPI is a high speed physical layer standard which provides 25 Mbps per line on a 32-bit parallel bus for a total bit rate of 800 Mbps. It is limited to 25 meters but in conjunction with fiber optic links, it can be extended to 2 kilometers. HPPI is also referred to as an ANSI standard X3T9.

Hub

A hub is a network component which centralizes circuit connections. Hubs started out as wiring concentrators, but have developed into sophisticated switching centers.

ICMP

ICMP stands for Internet Control Message Protocol. It is the portion of the Internet Protocol which handles network error control and diagnostic functions.

IEC

IEC stands for InterExchange Carrier. An IEC is a common carrier which carries user transmission across LATA boundaries.

Interoperability

Interoperability refers to the ability of different types of computers, networks, operating systems, and applications to work together effectively. An example of interoperability would be a TCP UNIX application using ASCII text files and exchanging data with an EBCDIC IBM host.

IP

IP stands for Internet Protocol. It is the IP in TCP/IP. IP is a network layer protocol and is responsible for getting a datagram through a network.

IPC

IPC stands for Interprocess Communication. It refers to the ability of a multitasking operating system to take advantage of processes available on other programs and/or computers. IPC uses Local Procedure Calls (LPC) to access information from a task running on the same computer. It uses Remote Procedure Call (RPC) to access information from a task running on another computer, usually across a network.

IPX

IPX stands for Internet Packet Exchange. It is Novell's network layer protocol, used for routing packets across a Novell NetWare LAN.

Isochronous Traffic

Isochronous traffic refers to the guarantee of bandwidth allocation for a given communications channel. This bandwidth is always available for the specified use. The need for isochronous channels arises when voice or real-time data communications is necessary. ATM and FDDI II are examples of technologies that provide isochronous services.

LAPB

LAP-B (or LAPB) stands for Link Access Protocol Balanced. It is an HDLC subset used primarily in X.25 communications. LAP-B is the Data Link layer sublayer of the X.25 protocol. LAP-B is responsible for point-to-point delivery of error-free frames. It is balanced because the LAP-B standard excludes the portions of the HDLC standard having to do with multidrop, "unbalanced" operation. It is very similar to LAP-D.

LAPD

LAPD or LAP-D stands for Link Access Procedure-D. It is part of the ISDN layered protocol. It is very similar to LAP-B. The D stands for "D channel." It defines the protocol used on the D channel to interface with the phone company's SS7 network for setting up calls and other signaling functions.

LAT

LAT stands for Local Area Transport. LAT is implemented by DECservers and VAXes or clusters to transfer character data via Ethernet. The term LAT also refers to the protocols used by nodes which support the architecture. One important purpose of the LAT architecture is to combine characters from several users into one Ethernet frame. This helps to utilize the Ethernet connection more fully. More importantly, it allows the host to handle input more efficiently, since the host can process several characters from several users each time it is interrupted for an Ethernet frame.

LATA

LATA stands for Local Access and Transport Area. After the breakup of ATT, the U.S. was partitioned into Local Access and Transport Areas (LATAs). LATA boundaries conform more or less to the standard metropolitan statistical areas defined by the U.S. Department of Commerce. Originally, they conformed to the boundaries of the areas served by the BOCs.

Latency

Latency is the amount of time that it takes a network component, such as a bridge or router, to transmit a received frame or packet.

Link

A link is a communications path between two or more communicating devices. Links are also referred to as channels, lines, circuits, and paths. Link Layer refers to the Data Link Layer or Layer 2 of the OSI model which is concerned with point-to-point communications.

LLC

LLC stands for Logical Link Control. The IEEE 802.2 standard defines the Logical Link Control (LLC) which is the upper layer of the 802 local area network protocol suite.

LocalTalk

LocalTalk refers to Apple Computer's 230K/bps cabling system and topology used to connect Apple computer products together.

Logical Channel

A logical channel is an X.25 term which means the logical connection (vs. physical connection) between the user terminal and the X.25 packet network. The DTE to DCE connection on both sides of "the cloud" has a logical channel number assigned by the software.

MAN

MAN stands for Metropolitan Area Network. MANs are being developed by data carriers to connect LANs in the same city. LANs, local area networks, extend across a single site and consist of one or more subnets, which are usually, but not necessarily, homogeneous. WANs are wide-area networks, often heterogeneous, which cover many sites and span large corporations and sometimes continents.

MAU

MAU stands for Media Access Unit and Multistation Access Unit, depending on the type of LAN involved. In the Token Ring world, it is a Multistation Access Unit. In the Ethernet world, it is a Medium (and sometimes media) Access Unit.

MAC

MAC stands for Media Access Control. It is one of the media specific IEEE 802 standards (802.3, 802.4, 802.5) which defines the protocol and frame formats for Ethernet, Token Bus, and Token Ring. It is the lower sublayer of the Data Link Layer of the OSI model.

MIB

MIB stands for Management Information Base. It is the standard for Network Management data for TCP/IP and Internet systems. It specifies the data that must be kept by a host or gateway in a network and the operations that may be performed on each.

Modem

Modem stands for Modulator/Demodulator. A modem is a device used to transmit computer data over the telephone network. It is a device which on the sending end converts digital data to an analog signal for transmission over an analog network (such as the telephone network). On the receiving end, the modem takes the analog signal and converts it back to digital.

Modulation

Modulation is the process of modifying the form of a carrier wave (electrical signal) so that it can carry intelligent information on some sort of communications medium. Digital computer signals (baseband) are converted to analog signals for transmission over analog facilities (such as the local loop). The opposite process, converting analog signal back into their original digital state, is referred to as demodulation. There are three basic types of modulation: frequency modulation, phase modulation, and amplitude modulation.

MOTIS

MOTIS stands for Message-oriented Text Interchange System. It is an application Layer message handling system developed jointly by the ISO and ITU-T.

Multidrop

Multidrop refers to a data communication configuration where multiple terminals, printers, and workstations are located on the same media and only one can communicate with the "master" at a given time. It is a form of unbalanced communication.

Multimode Optical Fiber

There are two basic types of fiber cables, multimode and singlemode. Light propagates along the fiber core in one of two ways depending on the type of material used. With multimode, the diameter of the core is large enough that the light reflects off the cladding as it travels down the core. With singlemode, the diameter of the core is small enough that light does not reflect and the signal is not altered as much as with multimode. Higher data rates and longer distances can be achieved with singlemode fiber.

Multiplexer

Computer equipment which allows multiple signals to travel over the same physical media is said to be a multiplexer. There are different types of multiplexers such as Time Division Multiplexers and Frequency Division Multiplexers.

Null Modem

A null modem is used to hook two computers together without the use of modems. Using a standard modem cable (RS232) from each of the two computers, you can configure them so that the two computers can communicate.

NAK

NAK stands for Negative Acknowledgment. In data communications, a node will send a NAK to another node to indicate that the frame should be retransmitted due to a problem during transmission or a problem at the receiver.

NAU

NAU stands for Network Addressable Unit. This is an IBM SNA term used to collectively describe PUs, LUs, and SSCPs. They are called Network Addressable Units because they have addresses and can communicate with one another. NAU Services are simply the services provided the NAUs in the nodes, together with the Transmission Control, Data Flow Control, and Function Management Layers of SNA. NAU Services include establishing connections, managing the end-to-end flow of data across the connections, providing certain application and user services, and overall management of the network.

NDIS

NDIS stands for Network Driver Interface Specification. NDIS was developed by Microsoft and 3Com to provide an interface between NIC drivers and networking protocols. The functionality of NDIS is comparable to ODI (Open Data-link Interface).

NetBEUI

NetBEUI stands for NetBIOS Extended User Interface. It is Microsoft's LAN Manager Transport Layer driver used in LAN Manager. It was developed by IBM in 1985 as a network transport protocol for LANs. In relation to the OSI model, NetBEUI is a network and transport layer protocol. NetBIOS and NetBEUI are integrated in the LAN environment to provide an efficient communications system.

NetBIOS

NetBIOS stands for Network Basic Input/Output System. It is a software system developed by Systek and IBM which has become the de facto standard for application interface to LANs. It operates at the Session Layer of the OSI protocol stack. Applications can call NetBIOS routines to carry out functions such as data transfer across a LAN.

NetWare Loadable Module

NetWare Loadable Module, or NLM, is software that runs on a Novell server. It is also commonly referred to as Communications Services software. It provides internetworking functions for Novell servers and workstations. The primary purpose for NLM is to allow database servers and communications servers to communicate with the server NOS. This is because applications, such as databases, cannot run on the server processor with NetWare.

Network Interface Card

A Network Interface Card (NIC) is any workstation or PC component (a hardware card) which allows the workstation or PC to connect and communicate to a network.

Network Operating System

The software that manages server operations and provides services to clients is called the Network Operating System (NOS). The NOS manages the interface between the network's underlying transport capabilities and the applications resident on the server.

ODI

ODI stands for Open Data-link Interface. It is a standard device driver from Novell which provides an interface between network hardware and the higher layer protocols. It supports multiple protocols and device drivers, allowing interconnectivity between various types of devices using different architectures and protocols.

OSI Defined

OSI stands for Open Systems Interconnection. OSI began as a reference model, that is, an abstract model for data communications, but now the model has been implemented and is in use in some data communications applications. The OSI model, consisting of seven layers, falls logically into two parts—layers one through four, the "lower" layers, are concerned with the communication of raw data, and layers five through seven, the "higher" layers, are concerned with the networking of applications.

Overhead

In data communications, overhead refers to the amount of data needed, in addition to the application data, to transmit information from one computer to another. In the example of the OSI model, this would be the header of each layer and the trailer of the Data Link Layer. The efficiency of communication devices is determined by the amount of overhead necessary to transmit user information. Efficiency typically decreases as the amount of overhead increases.

PABX

PABX stands for Private Automatic (or automated) Branch Exchange. It is the automated version of the PBX (Private Branch Exchange). Early PABXes were simply more sophisticated PBXes, providing additional voice services (such as call waiting) for analog extension loops. Today's PABXes are fully digital, not only offering very sophisticated voice services, such as voice messaging, but also integrating voice and data. A PABX can multiplex both voice and data onto T-carriers and can support both voice and data communications from one extension over a single pair of wires.

Packet

A packet is a unit of information that can be transmitted over a network. A packet is generated at the Network Layer of the OSI model protocol stack. Information contained in the header of a packet is sufficient to get the packet from the sending node to the receiving node, even when the packet must traverse through intermediate nodes. A packet can be an entire message or a segment of a much larger message generated at the application layer.

PAD

PAD stands for Packet Assembler/Disassembler. ITU-T developed a set of standards, informally called the Interactive Terminal Interface (ITI) standards, meant to provide access for terminals and DTEs that cannot execute the layers of X.25. The standards are X.3, X.28, and X.29. The ITI standards collectively define a "black box," called a packet assembler/disassembler, or PAD. A PAD "assembles" a stream of bytes originating from an asynchronous DTE (for example, from a personal computer) into X.25 packets and transmits them on the X.25 network. It performs the reverse operations for data sent back to the DTE.

Parity

Parity refers to the process of determining if data has been corrupted during asynchronous transmission of data. This is done by adding an additional bit (called a parity bit) to each character transmitted. There are two types of parity, odd parity and even parity. With even parity, the additional bit combined with the character makes the total number of ones an even number. With odd parity, the additional bit that is appended makes the total number of one bits odd. For example, if the ASCII character "A" is going to be transmitted, the seven bits that make up the "A" character 100 0001 will be transmitted. If even parity is used, the appended eighth bit will be a "0," keeping the total number of ones at two. If the parity is odd, the eighth bit will be a "1," making the total number of ones three.

PDN

PDN stands for Public Data Network. It is a network available to the public for transmission of data, usually using the X.25 protocol. Users dial into a server which allows access to the network. PDNs are especially popular in Europe.

PDU

PDU stands for Protocol Data Unit. The concept of PDU is used in the OSI reference model. From the perspective of a protocol layer, a PDU consists of information from the layer above plus the protocol information appended to the data by that layer.

Peer-to-Peer

Two programs or processes which use the same protocol to communicate and perform approximately the same function for their respective nodes are referred to as peer processes. With peer processes, in general, neither process controls the other, and the same protocol is used for data flowing in either direction. Communication between them is said to be "peer-to-peer."

Physical Address

In data communications, the term physical address refers to the address "burned in" to the hardware of a particular node. For instance, the physical address of a node connected to an Ethernet network is a 48-bit value which is unique for every Ethernet card purchased.

Ping

Ping stands for Packet Internet Groper. The ping command is used to determine if a destination host is reachable from the sender in an IP network. An ICMP echo packet is sent to a host and if a response is not sent back from that host within a specified amount of time, the host is considered "unreachable."

Polling

Polling is a method used to control communication between a master and slave node in an unbalanced data communication configuration. In an unbalanced configuration, the master "polls" the slave to ask if it has data to send or if it is in a state to receive data. An example of a protocol which uses a polling technique would be Synchronous Data Link Control (SDLC).

PPP

PPP stands for Point-to-Point protocol. PPP is a protocol that allows a computer to use the TCP/IP via a point-to-point link. PPP is based on the HDLC (High Level Data Link Control) standard which deals with LAN and WAN links and operates at the Data Link layer of the OSI model.

PPS

PPS stands for Packets Per Second. It is a measure of the number of data packets that can be transmitted over a given link during a one second interval. If a circuit is capable of transmitting data at a rate of 56,000 bps (bits per second) and the average packet size is 70 bytes (560 bits), the maximum PPS that can be transmitted is 100 in one second.

Presentation Layer

The Presentation Layer handles the representation of data as they flow between nodes. The lower layers provide the service of transferring data between nodes in an orderly fashion and ensuring that what is received is what has been sent. The Presentation Layer provides services that relate to the way data are represented.

Primary Rate Interface

The ISDN Primary Rate Interface is also called 23B+D. It is a circuit consisting of twenty-three 64Kbps channels for carrying voice, data, and video and a data channel used for carrying signaling information. It is similar to T1 signaling.

Print Server

A print server is a LAN based computer which provides users on the network access to a printer. The printer is therefore shared by multiple users.

Private Line

A private line is a telecommunications channel used exclusively by a single subscriber. Private lines are also referred to as dedicated lines or dedicated circuits.

Private Network

A private network is a network consisting of private lines, switching equipment, and other networking equipment that are provided for the exclusive use of one customer. In other words, the network and associated services of the network are not intended for usage by the general public. Most corporate networks are private networks.

Proprietary Network

A proprietary network is a network which is designed and implemented using nonstandards based equipment and protocols. Proprietary networks are typically created by purchasing equipment from one specific vendor. The opposite of a proprietary network would be an "Open" network which is based on standard protocols.

Protocol

Data communications involves the transfer of data between computer programs. Just as humans must share a common language in order to communicate, the programs must have a common protocol. The protocol simply defines the format and meaning of the data that the programs interchange.

PTT

PTT stands for Postal Telephone and Telegraph. These are telephone companies which are operated and regulated by government agencies. These are found in most countries other than the United States and Canada. In the United States, telephone service is offered by private, profit-making companies that are regulated by government agencies.

Public Network

A public network is a network which provides leased lines, packet switching services, and circuit switching services to the general public. It is the opposite of a private network. The term public switched network, however, usually refers to the voice telephone network.

Punchdown Box

A punchdown box is a passive device which is used to provide access to a group of wires for connectivity and testing purposes. Equipment such as telephones and computers are routed through a punchdown box before connecting to additional network components.

PVC

PVC stands for Permanent Virtual Circuit. There are two types of virtual circuits: permanent virtual circuits (PVCs) and switched virtual circuit (SVCs). A PVC behaves like a dedicated line between source and destination end-points. When activated, a PVC will always establish a path between these two end-points. It is usually used in the context of a packet switching (or cell switching) network.

RBOC

RBOC stands for Regional Bell Operating Company (also referred to as Regional Bell Holding Companies). There are seven different Regional Bell Operating Companies and twenty-two Bell Operating Companies. These are:

Ameritech: Bell, Indiana Bell, Michigan Bell, Ohio Bell, Wisconsin Bell

BellSouth: Southern Bell and South Central Bell

Bell Atlantic: Bell of Pennsylvania, C&P Telephone Companies of D.C., Maryland, Virginia and
West Virginia, Diamond State Telephone, and New Jersey Bell

Nynex: New England Telephone, New York Telephone

Pacific Telesis: Pacific Bell and Nevada Bell

Southwestern Bell: Southwestern Bell

US West: Mountain Bell Telephone, Northwestern Bell, and Pacific Northwest Bell

Remote Bridge

Bridges function at the Data Link Layer and can be local or remote. Local bridges connect networks in the same geographical area. Remote bridges use a telecommunication link (telephone line, satellite, etc.) and connect two or more LANs that are not located in the same geographical area.

Repeater

A repeater connects one cable segment of a LAN to others, possibly connecting differing media. For example, a repeater can connect thin Ethernet to thick Ethernet cables. It regenerates electrical signals from one segment of cable onto all of the others. Since it reproduces exactly what it receives, bit by bit, it also reproduces errors. But it is very fast and causes very little delay.

Response Time

Response time refers to the time it takes to receive a response once a request has been initiated. It is usually used in reference to interactive terminals requesting information from a host computer. An example would be the time it takes between the moment an enter key is pressed on a terminal until a full screen of data has been returned to that terminal. Factors that impact response time are link speed, protocol priority mechanisms, host processor utilization (how busy the host is), and network configuration.

RFC

RFC stands for Request For Comment. Request for Comment documents (RFCs) are working notes of the Internet research and development community. A document in this series may be on essentially any topic related to computer communication, and may be anything from a meeting report to the specification of a standard.

Router

Routers use the Network layer addressing to route information in the appropriate direction until a router recognizes the destination address. These types of hardware can relieve the various user nodes on the network of interface responsibilities. They perform the same basic function but with different responsibilities. They permit workstations on LANs to communicate over a wider area and maintain a higher level of performance.

Routing Table

A routing table is a data table stored in a computer's memory which contains the information necessary to route a frame, datagram, or packet to the next node in the communications path.

RPC

RPC stands for Remote Procedure Call. Remote procedure calls allow subroutines to be called remotely, that is, across the network.

SCSI

SCSI stands for Small Computer System Interface. It is a computer interface which essentially expands the 8-bit PC bus so that you can connect peripherals such as printers, hard drives, and floppy drives and exchange data at a rate of 5Mbps without stealing CPU cycles from the PC's main processor. Up to seven devices can be daisy-chained on a SCSI port. SCSI-2 is a 16-bit implementation of the SCSI bus.

Serial Communication

Serial communication refers to the transmission of bits which occur serially in time. Serial communication also implies that only a single communication channel is used. The asynchronous transmission of bits using the ASCII code set and a parity bit is a common method of serial communication.

Server

A server is a computer that can be shared by users of a Local Area Network. LAN servers are used for various purposes such as printing, file storage, and application program storage.

SIP

The internal SMDS protocols are called SMDS Interface Protocol-1 , -2, and -3 (SIP-1 through -3). They are a subset of IEEE 802.6. SMDS uses cell relay at layer 1, employing a cell with the same format as described for ATM; however, the format of the 5-byte header is different. The SIP protocols were defined for use by the phone companies.

SLIP

Serial communication over the telephone network has become much more important, and with this change has come the need to extend TCP/IP into the serial world of communications. Two protocols address this need. They are the Serial Line Interface Protocol (SLIP) and the Point-To-point Protocol (PPP). The SLIP protocol consists of two special characters, ESC (escape) and END. The END character (hex C0) marks the beginning and end of a SLIP frame. The ESC character (hex DB) is used to indicate where the data contains the ESC or END character so that the receiver will not interpret those occurrences as delimiters.

SMB

SMB stands for Server Message Block. It is the IBM PC LAN presentation layer protocol which is used to communicate with devices located on a LAN. It uses NetBIOS at the session layer to communicate across a LAN. Functions requiring LAN support, such as retrieving files from a file server, are translated into SMB commands before they are sent to the remote device.

SMDS

SMDS stands for Switched Multimegabit Data Service. SMDS is cell-oriented and uses the same format as the ITU-T B- ISDN standards. The ITU-T has standardized a connectionless Broadband Network Service (I.364). All of the RBOCs are conducting SMDS trials and plan to introduce tariffed SMDS service during the first half of 1995. However, Bell Atlantic is providing a pre-SMDS service. There is an equivalent interest by the European PTTs in an SMDS-type service.

SMTP

SMTP stands for Simple Mail Transfer Protocol. Unlike other communications protocols, which use binary codes in structured fields, SMTP uses plain English headers. SMTP defines a protocol and a set of processes that use the protocol to transfer e-mail messages between users' mailboxes. It does not define the programs used to store and retrieve mail messages.

SNAP

SNAP stands for Sub-Network Access Protocol. It is an LLC Header extension. It consists of five bytes of data, the first three bytes are referred to as the protocol ID and the last two bytes are referred to as the Ethertype.

SNMP

SNMP stands for Simple Network Management Protocol. It is based on the manager/agent model. Its primary purpose is to allow the manager and the agents to communicate. This protocol provides the structure for commands from the manager, notifies the manager of significant events from the agent, and responds to either the manager or agent. The original version of SNMP was derived from the Simple Gateway Monitoring Protocol (SGMP) and was published in 1988.

Socket

A socket is an API (Application Programming Interface) for interprocess communication between a UNIX based system and a TCP/IP network. A socket consists of either a TCP port and an IP address or a UDP port and an IP address. Application programs use sockets to communicate across a network to a peer process (TCP port).

SONET

SONET stands for Synchronous Optical NETwork. SONET standardizes optical transmission. The SONET standard defines a signal hierarchy. The basic building block is the STS-1 51.84 Mbps signal, chosen to accommodate a DS3 signal. The STS designation refers to the interface for electrical signals. The optical signal standards are correspondingly designated OC-1 (Optical Carrier-1), OC-2, etc.

SPX

SPX stands for Sequenced Packet Exchange. It provides transport layer functionality for the Novell NetWare System. The figure below shows the SPX format and how the SPX header fits into an Ethernet Frame.

SSCP

SSCP stands for System Services Control Point. The SSCP is an SNA host-based process which provides the services necessary to manage an SNA network.

StarLAN

The StarLAN is a proprietary Ethernet type Local Area Network developed and implemented by AT&T using telephone cabling and a star configuration.

STDM

STDM stands for Statistical Time Division Multiplexer. This type of multiplexer allows more terminals to be attached to a circuit than the capacity of the circuit can handle. It does this by taking into account that all devices do not transmit constantly at their maximum rates. A Stat MUX can be used instead of an ordinary multiplexer when voice channels are not required.

STP

STP stands for Shielded Twisted Pair. A twisted pair line, as the name implies, consists of a pair of wires twisted together. Because the wires are twisted together, electrical interference tends to affect both wires equally, so it does not affect the difference in the potential between the two wires. This makes twisted pair cable less susceptible to signal loss than if it were not twisted. Shielded twisted pair, like coaxial cable, is able to transfer data faster and over greater distances than unshielded twisted pair because of the additional protection from interference.

Subnet Mask

Internet (IP) addresses must all be unique. The IP address contains a network portion and a host portion. The host ID portion of the address must be unique for the particular network. Every node on the network must know how to tell which bits in the Internet address correspond to their physical network, or subnet. This is accomplished through a "subnet mask," which is set through the software in each node. If the subnet mask is incorrectly set by the user or System Administrator, the node will not be able to recognize its address in messages on the LAN and will not be able to communicate. The subnet mask, which must be consistent throughout the network, is a 32-bit hexadecimal word which "masks out" the node address.

Subnetwork

Networks can be classified according to the area over which they extend. A local area network (LAN) can consist of a few nodes up to several hundred but will typically be confined to a few buildings within a few thousand meters of one another. It can consist of subnetworks linked together in certain ways to form the larger, but still local, network.

Synchronous Transmission

With synchronous operation, modems must first closely synchronize their internal timing circuits (usually by transmitting a burst of bits of a feed length when the connection is established). To transmit data, the sending modem puts a one or a zero on the line every so often. The receiving modem samples the line on the same timetable and transmits the condition of the line (one or zero) to the DTE. They must stay in synchronization in order to communicate.

Tariff

A tariff is a document that describes the services offered by a carrier and defines the price for each service. If the services are regulated, then the tariff is filed with the regulating agency in order to place it in force, and of course the agency must review and approve it before it becomes effective.

TCP

TCP stands for Transmission Control Protocol. TCP is a transport layer protocol used to send messages reliably across a network. It is usually paired with the internet protocol (IP).

Telecommuting

Telecommuting refers to working in one's home instead of going to the office to work. Typically, it involves using some sort of communications link, such as a dial-up link, from your home to the office.

Teleconferencing

Teleconferencing refers to holding a conference between people who are located at different locations. It can be an audio link only or can be both audio and video.

Telephony

Telephony refers to the transmission of voice signals over a distance (i.e. using telephone equipment such as switches, telephones, and transmission media).

TELNET

Telnet is the TCP standard protocol for connection of remote terminals to host nodes. TELNET is a virtual terminal protocol. Virtually all nongraphical applications written for the TCP/IP environment use the TELNET protocol for input and output to the user.

Terminal Server

A terminal server is a device which allows multiple terminals or workstations to be multiplexed onto a LAN. Each port of a terminal server can connect a local or remote terminal (or a PC which is emulating a terminal) to any of the services on the Ethernet. An example of a terminal server is a DECserver which combines characters from several users into one Ethernet frame for more efficient use of the LAN media.

Thicknet

Thicknet, also known as 10Base 5 and Yellow Wire, can carry a signal for 500 meters before a repeater is required. The maximum number of nodes allowable in a trunk segment is 100. The maximum number of trunk segments allowable in an Ethernet network is five, of which only three may be populated with nodes.

Thinnet

Thinnet, or 10Base2, can carry a signal for 185 meters before a repeater is required. The maximum number of nodes which can be connected to a thinnet trunk segment is thirty. Thinnet is also called "cheapernet" because it is the cheapest way to wire an Ethernet network.

Throughput

Throughput refers to the total amount of data which can be transmitted from source to destination over a given amount of time. It does not include the overhead associated with protocol headers and trailers or network delays.

Time Division Multiplexing

TDM stands for Time Division Multiplexing. This type of multiplexing combines many digital bit streams with relatively low bit frequency into a single bit stream with a relatively high bit frequency. It is, in essence, a way for many slow communications channels to "time share" a very fast channel. The advantage, of course, is that the cost per bit transmitted on a single fast channel is lower than on slower channels.

TN3270

TN3270 is a shortened version of Telnet 3270. TN3270 is a TCP/IP application which delivers 3270 datastream via TELNET.

Topology

Topology refers to the specific physical configuration of a network or a portion of a network. Ring and Star are examples of different network topologies.

Transport Layer

The Transport Layer is the fourth layer in the OSI Reference Model. It is concerned with controlling the transport of messages in a network.

UART

UART stands for Universal Asynchronous Receiver/Transmitter. A UART is a chip located on a circuit board in a computer. This circuit board also contains the serial port which communicates with a serial device such as a modem or a serial printer. The job of the UART is to convert serial data into parallel data and vice versa for the computer. The internal computer bus is a parallel bus and needs the UART to convert this parallel data to asynchronous serial data before it can be transmitted to a modem or other device via a serial cable such as RS232. It is also responsible for taking in asynchronous serial data from a modem or other device and converting it to parallel data for the computer bus.

Unshielded Twisted Pair

Unshielded Twisted Pair (UTP) can carry a signal 100 meters before a repeater is necessary. Unlike Thinnet or Thicknet, a hub or concentrator is also required. The maximum number of nodes per segment is 1024, which is also the same number of nodes that may exist on an entire network based on lOBaseT.

Value Added Network

A Value Added Network (VAN) is a data communications service (or any networking service) with added functionality (or value), provided to users for a charge. An example of a VAN would be a public X.25 network where services are provided such as guaranteed response times or other "high quality services."

Virtual Circuit

A virtual circuit is a communications path that appears to be a single circuit even though the data may take varying routes between the source and destination nodes. This concept has its roots in X.25 packet-switching. There are two types of virtual circuits: permanent virtual circuits (PVCs) and switched virtual circuit (SVCs).

Virtual Terminal

As networks developed, many different terminals were in use, so to avoid the problem of incompatibilities, the concept of the virtual terminal was developed. Virtual terminal protocols provide an abstract definition of a terminal that is specific in terms of the control codes required to produce a specific result, but is also general in nature. The virtual terminal definition combines all of the common features of most real terminals but leaves out the "bells and whistles" that make terminals unique to their vendor.

VTAM

VTAM stands for Virtual Telecommunications Access Method. VTAM is an IBM host computer program which controls the interaction of terminal traffic to the host for IBM SNA networks.

WAIS

WAIS stands for Wide Area Information Server. It is an system used to look up information which is available on the Internet.

WAN

WAN stands for Wide Area Network. Wide area networks are essentially interconnected LANs or MANs. They can be homogeneous, interconnecting like networks, but are often heterogeneous, that is, interconnecting LANs or MANs that have been built using different technologies. A WAN can span campuses, cities, states, or even continents. Typically, only one node on each LAN or MAN, called a gateway, connects to the WAN. Other nodes communicate with the WAN via the gateway.

World Wide Web

The World Wide Web is a public domain hypertext application which is used for finding and retrieving information on the Internet.

X Modem

The X Modem protocol is a data communications file transfer protocol developed for transmitting files between PCs.

Y Modem

Y Modem is a data communications file transfer protocol similar to X Modem. The basic difference between Y Modem and X Modem is that the block size was increased from 128-byte blocks to 1024-byte blocks with the Y Modem protocol. It is compatible with the X Modem protocol when sending 128-byte blocks.

Z Modem

The Z Modem protocol is a data communications file transfer protocol developed for transmitting files between PCs. It is similar to both the X Modem and Y Modem protocols. Z Modem, however, is implemented using full duplex communication and does not require each block of data to be acknowledged as does the X Modem and Y Modem file transfer protocols. Z Modem continues to send blocks of data until a negative acknowledgment is received (NAK). It then retransmits all blocks prior to the block which was NAKed.

APPENDIX A - LIST OF ACRONYMS

A

AARP: Appletalk Address Resolution Protocol
AAUI: Apple Attachment Unit Interface
ACF: Advanced Communication Facility
ACL: Access Control List
ADB: Apple Desktop Bus
ADSL: Asymmetrical Digital Subscriber Line
ADSP: Appletalk Data Stream Protocol
ADDCP: Advanced Data Communication Control Procedures
ADPCM: Adaptive Differential Pulse Code Modulation
AEA: ASCII Emulation Adapter
AEP: Appletalk Echo Protocol
AFP: Appletalk Filing Protocol
AIX: Advanced Interactive eXecutive
ALAP: Appletalk Link Access Protocol
ALOE: AppleTalk Low Overhead Encapsulation
AM: Amplitude Modulation
AMI: Alternate Mark Inversion
ANSI: American National Standards Interface
AOCE: Apple Open Collaborative Environment
AOL: American On-Line
APAD: Asynchronous Packet Assembler Disassembler
API: Application Program Interface
APPC: Advanced Program-to-Program Communication
APPN: Advanced Peer-to-Peer Networking
ARA: Apple Remote Access

ARP: Address Resolution Protocol
ARPA: Advanced Research Projects Agency
ASCII: American Standard Code for Information Interchange
ASN: Abstract Syntax Notation
ASP: Appletalk Session Protocol
ATM: Asynchronous Transfer Mode
ATP: AppleTalk Protocol
AURP: Appletalk Update-based Routing Protocol
AUI: Attachment Unit Interface
AWG: American Wire Gauge

B

BBS: Bulletin Board System
BCD: Binary Coded Decimal
BER: Bit Error Rate
BERT: Bit Error Rate Tester
BGP: Border Gateway Protocol
BIOS: Basic Input Output System
BISDN: Broadband Integrated Services Digital Network
BIT: Binary digIT
BIU: Basic Information Unit
BLU: Basic Link Unit
BOC: Bell Operating Company
BPS: Bits Per Second
BRI: Basic Rate Interface

C

CAD: Computer Aided Design
CBR: Committed Burst Rate
CCITT: Consultative Committee for International Telegraph and Telephone
CCL: Connection Control Language
CDDI: Copper Data Distributed Interface
CDIS: Custom Defined Integrated Services
CGI: Common Gateway Interface
CGP: Core Gateway Protocol
CIR: Committed Information Rate
CISC: Complex Instruction Set Computer
CIX: Commercial Internet eXchange

CLNP: ConnectionLess Network Protocol
CMIP: Common Management Information Protocol
CMIS: Common Management Information Service
CO: Central Office
COM: COMmercial
COM: Common Object Module
COS: Corporation for Open Systems
CPE: Customer Premise Equipment
CPU: Central Processing Unit
CRC: Cyclical Redundancy Check
CSMA/CD: Carrier Sense Multiple Access with Collision Detection
CSU: Channel Service Unit
CTS: Clear To Send

D

DAL: Database Access Language
DASD: Direct Access Storage Device
DBMS: DataBase Management System
DCE: Data Communication Equipment
DCE: Distributed Computing Environment
DDP: Datagram Delivery Protocol
DDS: Digital Data Service
DNA: Digital Networking Architecture
DEC: Digital Equipment Corporation
DES: Data Encryption Standard
DLCI: Data Link Connection Identifier
DME: Distributed Management Environment
DNA: DEC Network Architecture
DOS: Disk Operating System
DPI: Dot Per Inch
DPSK: Differential Phase Shift Keying
DSP: Digital Signal Processors
DQDB: Distributed Queue Dual Bus
DSAP: Destination Service Access Point
DSE: Data communications Switching Equipment
DSU: Digital Service Unit
DTE: Data Terminal Equipment

E

EDI: Electronic Data Interchange
EDU: EDUcation
EFS: Error Free Seconds
EGP: Exterior Gateway Protocol
EIA: Electronic Industry Association
ELAP: Ethernet Link Access Protocol
EMI: ElectroMagnetic Interference
EMIF: ESCON Multiple Image Facility
EOF: End Of File
EOT: End Of Text
EPS: Encapsulated PostScript
ESCON: Enterprise Systems CONnectivity
ESCM: ESCon Manager
ESD: Electrostatic Discharge
ESD: Electronic Software Distribution
ESF: Extended SuperFrame

F

FAQ: Frequently Asked Questions
FC: Fibre Channel
FDDI: Fiber Distributed Data Interface
FEP: Front End Processor
FTP: File Transfer Protocol

G

GIF: Graphics Interchange Format
GOSIP: Government OSI Profile
GOV: GOVernment
GUI: Graphical User Interface

H

HCSS: High Capacity Storage System
HDLC: High-level Data Link Control
HDSL: High bit rate Digital Subscriber Line
HDX: Half DupleX

HFS: Hierarchical Filing System
HIPPI: HIgh Performance Parallel Interface
HMI: Hub Management Interface
HPFS: High Performance File System
HREF: Hypertext REFerence
HSSI: HIgh Speed Serial Interface
HTML: HyperText Markup Language
HTTP: HypeText Transport Language

I

IAB: Internet Architecture Board
IBM: International Business Machines
ICMP: Internet Control Message Protocol
IDE: Intergrated Drive Electronics
IDP: Internet Datagram Protocol
IDRP: InterDomain Routing Protocol
IEC: InterExchange Carrier
IETF: Internet Engineering Task Force
IMAP: Interactive Mail Access Protocol
IP: Internet Protocol
IPC: InterProcess Communication
IPX: Internet Package Exchange
IRC: Internet Relay Chat
IRQ: Interrupt ReQuest
ISDN: Integrated Services Digital Network
ISO: International Organization for Standardization
IXC: IntereXchange Carrier

J

JCL: Job Control Language
JTM: Job Transfer and Management

L

LAN: Local Area Network
LAP: Link Access Protocol
LAPB: Link Access Protocol Balanced
LAPD: Link Access Protocol D-channel
LAT: Local Area Transport

LATA: Local Area and Transport Access
LAVC: Local Area Vax Cluster
LAWN: Local Area Wireless Network
LEC: Local Exchange Carrier
LLAP: Localtalk Link Access Protocol
LLC: Logical Link Control
LSL: Link Support Layer
LU: Logical Unit
LXC: Local eXchange Carrier

M

MAC: Media Access Control
MAN: Metropolitan Area Network
MAP: Manufacturing Automation Protocol
MAPI: Message Application Programming Interface
MAU: Multistation Access Unit
MB: MegaByte
MBPS: Million Bits Per Second
MHS: Message Handling Service
MIME: Multipurpose Internet Mail Extension
MIP: Million Instructions per Second
MLID: Multiple Link Interface Driver
MOP: Maintenance Operations Procedure
MOTIS: Message Oriented Text Interchange System
MS-DOS: MicroSoft Disk Operating System
MTA: Message Transfer Agent
MTS: Message Transfer System
MTU: Maximum Transmission (Transfer) Unit

N

NAK: Negative AcKnowledgement
NATA: National American Telephone Association
NAU: Network Addressable Unit
NBP: Name Binding Protocol
NCB: Network Control Block
NCCF: Network Communications Control Facility
NCP: Netware Core Protocol
NCP: Network Control Program
NDIS: Network Driver Interface Specification

NET: NETwork
NFS: Network File System
NIC: Network Interface Card
NIU: Network interface Unit
NLDM: Network Logical Domain Manager
NLM: Netware Loadable Module
NOS: Network Operating System
NPDA: Network Problem Determination Manager
NREN: National Research and Educational Network

O

OC: Optical Carrier
OCR: Optical Character Recognition
ODA: Open Document Architecture
ODI: Open Data link Interface
OLE: Object Linking and Embedding
OLTP: On-Line Transaction Processing
OOP: Object-Oriented Programming
ORG: ORGanization
OS: Operating System
OSI: Open Systems Interconnection
OSPF: Open Shortest Path First

P

PABX: Private Automated Branch eXchange
PAD: Packet Assembler/Disassembler
PAP: Printer Access Protocol
PBX: Private Branch eXchange
PC: Personal Computer
PCI: Peripheral Component Interconnect
PCN: Path Control Network
PDN: Public Data Network
PEP: Packet Exchange Protocol
PIU: Path Information Unit
PLP: Packet Layer Protocol
POP: Post Office Protocol
POP: Point Of Presence
PPS: Packets Per Second
PPP: Point-to-Point Protocol

PU: Physical Unit
PVC: Permanent Virtual Circuit

R

RAID: Redundant Array of Inexpensive Disks
RAM: Random Access Memory
RARP: Reverse Address Resolution Protocol
RBHC: Regional Bell Holding Company
RBOC: Regional Bell Operating Company
RCP: Remote Communications Processor
RIP: Routing Information Protocol
RISC: Reduced Instruction Set Computing
RNR: Receiver Not Ready
ROM: Read Only Memory
RPC: Remote Procedure Call
RTMP: Routing Table Maintenance Protocol
RTS: Ready To Send
RU: Request Unit
RU: Response Unit

S

SABM: Set Asynchronous Balanced Mode
SAP: Service Access Point
SAR: Segmentation And Reassembly
SCC: Serial Communications Controller
SCSI: Small Computer System Interface
SDU: Service Data Unit
SDLC: Synchronous Data Link Control
SFT: System Fault Tolerant
SIP: SMDS Interface Protocol
SGML: Standard Generalized Markup Language
SGMP: Simple Gateway Monitoring Protocol
SIMM: Single Inline Memory Module
SLIP: Serial Line Internet Protocol
SMB: Server Message Block
SMDS: Switched Multimegabit Data Service
SMTP: Simple Mail Transfer Protocol
SNA: Systems Network Architecture
SNAP: Sub-Network Access Protocol
SNMP: Simple Network Management Protocol

SNRM: Set Normal Response Mode
SONET: Synchronous Optical NETwork
SOF: Start of Frame
SOFD: Start Of Frame Delimiter
SPP: Sequenced Packet Protocol
SPX: Sequenced Packet eXchange
SSA: Serial Storage Architecture
SSAP: Source Service Access Point
SSCP: System Services Control Point

T

TCP: Transmission Control Protocol
TCP/IP: Transmission Control Protocol/Internet Protocol
TDM: Time Division Multiplexing
TIC: Token ring Interface Card
TLA: Three Letter Acronym
TLI: Transport Layer Interface
TOP: Technical Office Protocol

U

UA: Unknown Acknowledgement
UART: Universal Asynchronous Receiver/Transmitter
UDP: User Datagram Protocol
UHF: Ultra-High Frequency
UPS: Uninterruptible Power Supply
URL: Uniform Resource Locator
USRT: Universal Synchronous Receiver/Transmitter
UTP: Unshielded Twisted Pair
UUCP: Unix-to-Unix Copy Program

V

VAX: Virtual Address Extension
VCC: Virtual Channel Connection
VCI: Virtual Channel Interface
VIP: Vines Internet Protocol
VRML: Virtual Reality Markup Language
VTAM: Virtual Telecommunications Access Method

W

WAIS: Wide Area Information Servers
WAN: Wide Area Network
WLAN: Wireless Local Area Network
WWW: World Wide Web

X

XDF: eXtended Distance Fiber
XID: eXchange ID
XNS: Xerox Network System

Z

ZIP: Zone Information Protocol
ZIT: Zone Information Table

APPENDIX B - USING THE CD-ROM

INTRODUCTION

Copyright Notice
Copyright © 1996 WestNet Inc.
All Rights Reserved

No part of this software may be reproduced or transmitted in any form without the express written consent of WestNet. The software furnished with this book is a *Single User* copy. For more information about other products, services or multiuser packages, please contact us at the following location:

WestNet
6999 Poppy Court
Arvada, Co 80007
(303) 424-9168 Fax/Phone

Welcome to DNH Software, an easy-to-use computer networking tutorial supplied with the Data Network Handbook book. DNH Software is designed to enable you to learn the technical details of networking using your computer. It is also a comprehensive network reference, containing a thorough glossary and information on thousands of networking topics.

DNH Software runs on both Microsoft Windows based personal computers as well as on Apple Macintosh computers. It utilizes the latest in Windows technology. This allows you great flexibility in using the software. You can read through the tutorial sequentially by topic, or search for words and topics. You may browse forward or backward from any point in the system.

DNH Software comes complete with over 600 hundred diagrams which can be viewed in context with the appropriate topics. Each topic, definition and diagram can be printed to your printer.

DNH Software is designed for computer users wanting to learn the details of networking technology from both a technical and non-technical perspective. The system allows you to decide how fast or slow you want to learn the material and how in-depth you want to go on a particular topic.

System Requirements

To run DNH Software, you need:

- A Personal Computer running Microsoft Windows version 3.1 or later.

- A Macintosh running System Software version 6.0.5 or later.

- 20 Megabytes of hard disk space (this assumes you are running DNH Software from your hard drive).

To get the most out of DNH Software, a VGA color monitor with at least 640 X 480 resolution is recommended.

If you want to print text, a Windows or Macintosh compatible text or graphics printer is required. To print diagrams, a Windows or Macintosh compatible graphics printer is required.

Installing DNH Software

Installation on a Mac

To install DNH Software on a Macintosh computer, insert the CD-ROM into the appropriate drive. Click on the "VISE" icon (VISE is the installer program). Click on install and follow the instructions.

Installation on a PC

To install DNH Software on a Personal Computer, complete the following steps:

Note: This example installation is taken from Windows 3.11. When DNH Software is being used under Windows NT or Windows 95, installation is slightly different.

Installation Instructions

1. Start Microsoft Windows on you computer.
2. Insert DNH CD into the the CD-ROM drive.
3. From Program Manager:
    a. Select File Menu
    b. Select Run.
4. In the Run dialog box (See Figure 1), type a:\setup

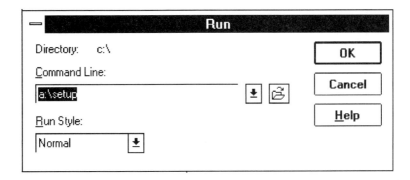

Figure 1. Run Dialog Box

Note: if you are running the setup program from a drive other than drive a:, then use that drive letter in place of the letter "a" in the command above.

5. Click the OK button and the Setup Dialog Box will appear (See Figure 2).
6. Click the OK in the Destination for DNH Software dialog to accept the default directory (C:\BASICS in this example), or enter the desired destination directory in the dialog.
7. You will be prompted if additional information is needed.

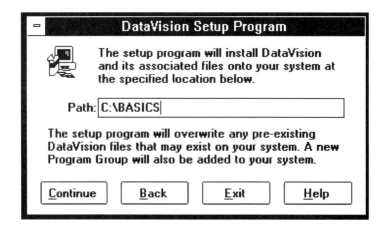

Figure 2. Setup Dialog Box

Using DNH Software

To run DNH Software (Windows PC version):
1. Start Microsoft Windows
2. Double click the Program Manager group labeled "DNH Software".
3. Double click the DNH Software icon.

To run DNH Software (Mac version):
1. Start the Macintosh.
2. Double click the DNH Software Icon.

Details of Operation

DNH Software uses the Microsoft Windows Help system provided with the Microsoft Windows operating system. To get the most out of DNH Software, you should be familiar with running Windows. If you are new to Windows, refer to your Windows documentation.

Starting DNH Software

To run DNH Software (using a PC):
1. Start Windows
2. From Program Manager, double click the DNH Software group icon.
3. Double click the DNH Software program icon to run DNH Software. The DNH Software main window (Contents Screen) appears (See Figure 3):

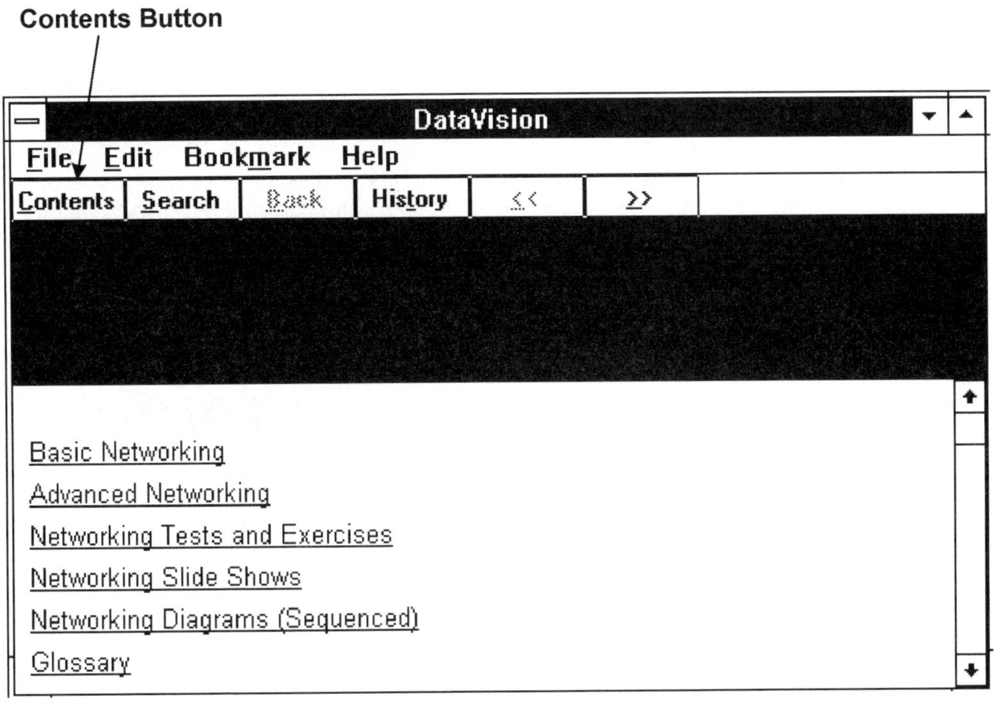

Contents Button

Figure 3. Contents Screen

Contents Screen

The contents screen, shown above, is displayed when DNH Software starts. This contents screen is an interactive main index to the topics contained in the system. Click any green topic text to view that topic. Click the Contents button on the button bar at any time to return to the contents screen. The actual contents of the version you are running may vary from the one pictured above. This is because this software is updated quarterly with new information on networking.

Navigating through Topics

After you select a topic from the contents screen, the main window for that topic appears. A topic window may contain another list of several more subtopics, any of which can be selected by clicking the mouse. The number of levels for each topic varies, depending on the subject. The following attributes are common among the topic and subtopic windows (Refer to Figure 4).

*   The scroll bar, on the right side of the window, allows you to scroll through the topic text one screen at a time. Click in the scroll area to move through topic text.

*   The maximize button, on the upper-right corner of the window, allows you to enlarge the window's viewing area. Click the button again to return the window to its original size and position.

Use the maximize button in the diagram window to enlarge the diagram window size. Scroll bars may be available, depending on the size of the diagram, allowing you to scroll the diagram window to view different parts of the diagram.

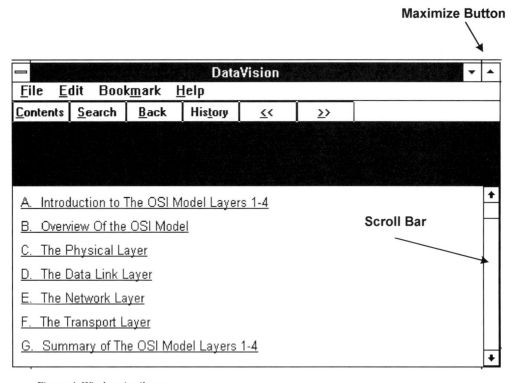

Figure 4. Window Attributes

The default view contains a text window (main window) and a diagram window. When both windows are visible, the screen will look like that shown in Figure 5.

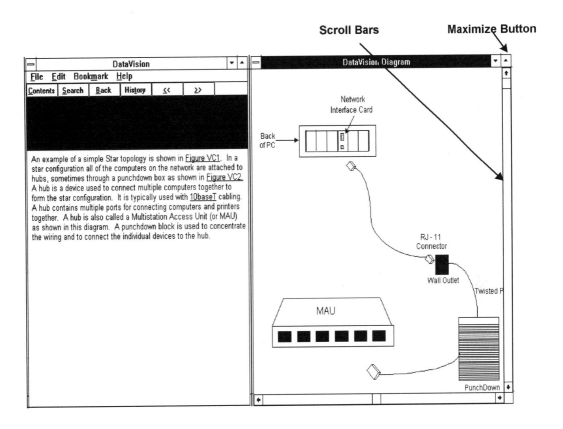

Figure 5. Default View (Multiple Windows)

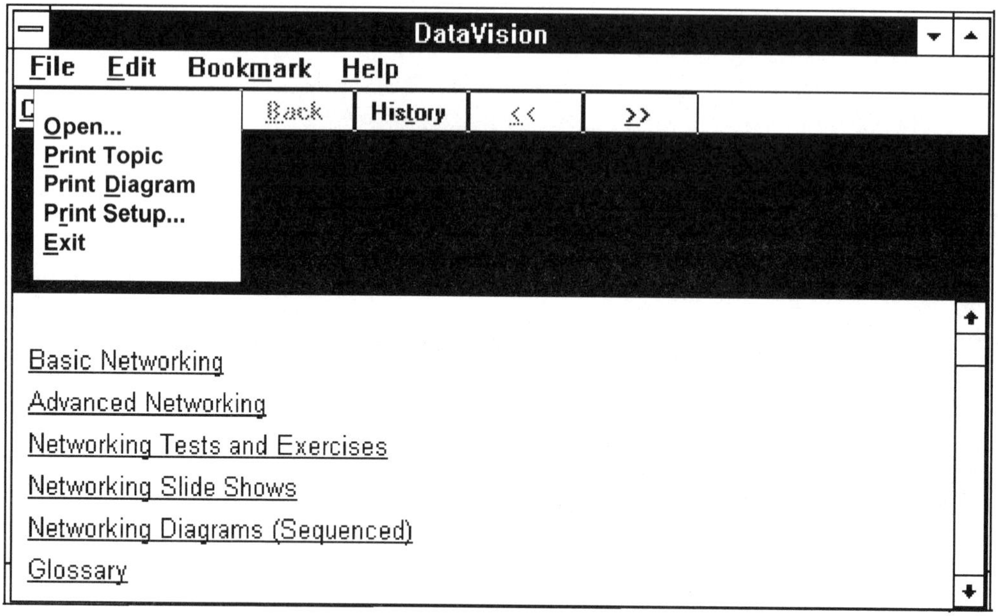

Figure 6. Printing Topics and Diagrams

Printing Topics and Diagrams

To print the current topic, select Print Topic from the menu. To print the current diagram, select Print Diagram from the menu (See Figure 6). Printer selection and setup is done via the Printer Setup menu.

Quitting DNH Software

To quit DNH Software, select Exit from the File menu as shown in Figure 6.

DNH Software as a Tutorial

To use DNH Software as a tutorial, you view each topic in order. To do this, click the >> (forward) button to setup through each topic. This allows you to view each topic and sub-topic in order. When diagrams are available, highlighted text will allow to choose the appropriate diagram.

When you end a session with DNH Software, use the bookmark menu option to set a bookmark. This saves your place in the tutorial and allows you to return quickly upon starting your next tutorial session (See Figure 7).

Searching for words and topics

DNH Software can search for words and topics. To search, click the Search button on the button bar. The Search dialog appears:

To search for a word or topic (Refer to Figure 8):

1. Type the search word or topic in the edit control, or select it from the list box.
2. Click the Show Topics button.
3. Select a topic from the list displayed in the lower portion of the dialog.
4. Click the Go To button.

DNH software displays the selected topic in the main window:

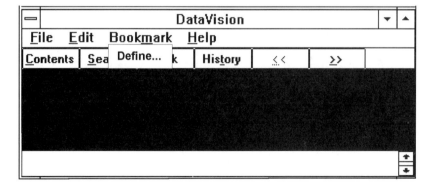

Figure 7. Defining Bookmarks

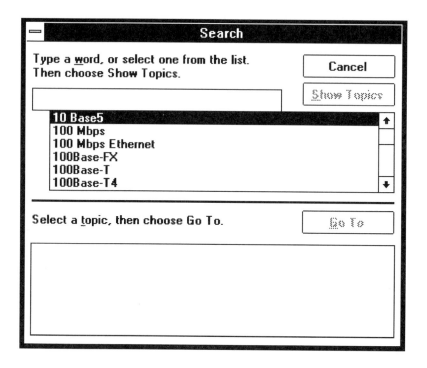

Figure 8. Search Window

Button Bar and Menu Reference

The Button Bar is located just below the DNH Software menu bar (See Figure 9). The buttons and their functions are listed below:

Contents: View DNH Software's main contents.

Search: Activate Search dialog to search for words and topics.

Back: Return to the previously viewed topic. Retraces your path through DNH Software.

History: View complete, sequential list of all DNH Software topics you have viewed during the current session.

<<: View the previous topic. When no previous topics are available, this button is dimmed.

>>: View the next topic. When no more topics are available, this button is dimmed

Button Bar

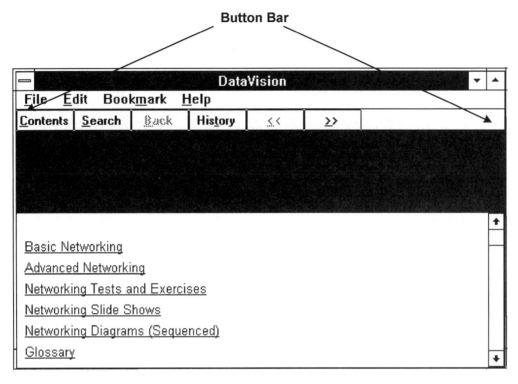

Figure 9. Button Bar

Troubleshooting

DNH Software is designed to be as simple to use as possible. However, should you have difficulty running the software, please review the following before calling WestNet.

Problem
When attempting to run DNH Software, a dialog appears with the message "Cannot open Help
file."

Solution
Microsoft Windows cannot find the *.HLP file, which is the help file containing DNH Software
data.

To solve this problem:

1. Determine the correct path to the *.HLP file (if you accepted the default path on setup, the path is "C:\BASICS").

2. From the Windows Program Manger, edit the properties field for DNH Software as follows:

> a. Select the DNH Software icon by single clicking the icon.
> b. Select "Properties" from the "File" Menu.
> c. Make sure the correct path, as determined above, is entered in the "Working Directory" edit box of the Program Item Properties dialog.

Problem
When attempting to print a diagram from DNH Software, a dialog appears with the message "Routine Not Found."

Solution
Microsoft Windows cannot find the PNT001.DLL file, which is a file containing program routines used by DNH Software.

To solve this problem:

1. Determine that the PNT001.DLL file is located in the same directory as DATAVZN.HLP (if you accepted the default directory on setup, this file should be located in the directory "C:\BASICS").

2. If so, set the Working Directory path to the proper directory, as discussed in the previous section.

3. If not, reinstall DNH Software.

Problem
When attempting to print a diagram from DNH Software using Windows95, a dialog box appears which says, "cannot find pnt.hlp file".

Solution
Microsoft Windows cannot find one of the necssary files.

To solve this problem:

1. Determine where the .hlp file is located.

2. Copy this file and name it "pnt.hlp".

If you have other problems, contact WestNet.

Contacting WestNet

Call:
   Fax and Voice: 303-424-9168

Write:
   WestNet
   6999 Poppy Court
   Arvada, Colorado 80007

INDEX

CD Operating Instructions

To run the software you will need:

1. IBM PC or compatible running any one of the following:
   - Microsoft Windows version 3.1 or higher
   - Windows 95
   - Windows NT
   - Windows for Workgroups

   OR

   Macintosh running system 7.0.0 or higher

2. CD ROM Drive

3. 20 Megabytes of hard drive space unless running from the CD